FIRE BLIGHT

The Foundation of Phytobacteriology

Selected Papers of

Thomas J. Burrill, Joseph C. Arthur, and Merton B. Waite

Edited, with introductory and concluding essays, by

Clay S. Griffith, Turner B. Sutton, and Paul D. Peterson

APS PRESS
The American Phytopathological Society
St. Paul, Minnesota

This book has been reproduced directly from pages submitted in final form to APS Press by the editors of the volume. No editing or proofreading has been done by the Press.

Cover photographs of T. J. Burrill, J. C. Arthur, and M. B. Waite courtesy of the National Archives; photograph of Arthur's laboratory notebook courtesy of the New York State Agricultural Experiment Station, Geneva

Library of Congress Control Number: 2003105599
International Standard Book Number: 0-89054-309-7

Printed in the United States of America on acid-free paper

The American Phytopathological Society
3340 Pilot Knob Road
St. Paul, Minnesota 55121-2097, USA

Contents

Preface

This Phytopathological Classic is a collection of research papers that document the essential work on fire blight from the late 19th and early 20th centuries. The authors of these papers are three important scientists who did pioneer work on fire blight – Thomas Jonathan Burrill, Joseph Charles Arthur, and Merton Benway Waite. It was these scientists who first revealed that a bacterium, in this case *Erwinia amylovora*, could cause a disease in plants and that insects could serve as vectors for its spread. Their discoveries belong to the same class of scientific breakthroughs as the work on rust and smut disease in the 1850s, which revealed the fact of fungal pathogenicity, and the research on tobacco mosaic disease in the first years of the 20th century, which established the causal role of viruses in plant disease. The unique place of fire blight in the development of plant pathology is sufficient reason, in and of itself, for this research to be considered a formative event in the science of phytopathology and would provide ample justification for the collection of papers that compose this volume.

There is, however, more about the history of fire blight research that makes it worthy of attention. The importance of early studies of fire blight touches on the development of biology and bacteriology. The revelation that bacteria could cause plant disease came at virtually the same time as the similar discovery with human and animal diseases. And the research in the role of insects in the spread of the disease appeared contemporaneously with reports of insect vectors in animal disease. Beyond that, the early research on this disease provides an excellent vehicle for viewing the development of plant pathology, whether in terms of concepts, personalities, or institutions, during the formative years of the science.

The papers collected in this volume have been chosen by the editors to tell the story of the exciting first decades of fire blight research. But this does not necessarily mean we chose the most famous papers on the topic. There were times when we eschewed better known papers for those that were less well known but, to our minds, more appropriate. Often the better known papers fell into this category simply because they were published in more popular or visible sources. In some cases, papers that were printed in more obscure publications may actually contain greater detail on the disease, the need for research, and on the author's experimental techniques. As such, if certain papers appeared to better reveal the developing nature of plant pathology, we selected them for this collection.

Our editorial procedures need some brief discussion. The enclosed papers are not represented as facsimiles of the originals; they have been retyped to create visual standardization for ease of reading. The layouts, however, have been retained as closely as possible to the originals, such as the placement of figures and tables. We have also retained grammar, spelling, and capitalization conventions. Misspellings have not been corrected, but have been called out in the text. We have also retained all footnotes from the originals, but we have chosen to alter their format to standard consecutive numerals running throughout the paper rather than use the authors' original methods which were either renumbering the footnotes on each page or designating notes with repeating symbols rather than numerals.

The editors would like to thank a number of people who assisted in the preparation and production of this volume. First of all, the staffs at the National Archives, the National Agricultural Library, and the Systematic Botany and Mycological Laboratory were considerably helpful to our research. We also received enormous assistance from the librarians and archivists at Cornell University, Iowa State University, and the University of Illinois, as well as the American Association for the Advancement of Science and the Philadelphia Academy of Natural Sciences. We particularly would like to thank Alan Jones for his perceptive comments on the manuscript. The technical elements of production were made possible by the expertise and patience of Patricia Sullivan and Sandra Townes. Lastly, as always, we express our appreciation to the staff of APS Press, with special gratitude to Karen-Beth Scholthof for shepherding this project through the process.

Introduction
Fire Blight: The History, Science, and Politics of a Disease

There is perhaps no disease of fruit trees so utterly devastating as fire blight. This disease plagues pear, apple, and other pome fruits as well as many related ornamentals such as *Crataegus* and *Pyracantha*. The term fire blight is a fitting description because flowers, twigs, and branches that have been invaded by the bacterium appear to be scorched.[1] The disease can destroy orchards in a single season, leaving trees blackened as if by wildfire, and its presence effectively limits the range in which certain apple and pear cultivars can be grown economically.

Fire blight is caused by the bacterium *Erwinia amylovora* (Burrill) Winslow et al. This bacterium is a unique and fascinating pathogen. Even nearly 150 years after its discovery as the cause of fire blight, there are still many questions regarding *E. amylovora* and how it causes disease. We still do not fully understand the nature of its pathogenicity, why it is the only species of bacteria that causes fire blight, nor why it causes fire blight only on certain species of *Rosaceae*. In addition, other issues such as the rapidity of its movement in the host has been noted as remarkable, as well as its ability to move through trees without producing any tissue-dissolving enzymes.[2] It has been noted that "*E. amylovora* can be considered both as a well-known bacterial species and as a poorly known bacterial plant pathogen."[3]

E. amylovora and fire blight are important to the history of plant pathology and microbiology. Fire blight was the first plant disease found to be caused by a bacterium and opened the way for all future study of bacterial plant diseases and phytobacteriology in general. This momentous event occurred during the same period of the 19th century when other scientists were uncovering the role of bacteria in the diseases of humans and animals, sparking the explosive development of microbiology. The plant side of these exciting days has been lost in the glare of the work of microbiological giants like Pasteur and Koch, but it deserves its place in scientific history. In addition, fire blight was also the disease under consideration when the role of insects as vectors of plant pathogens was proven. Again, this landmark discovery occurred in the shadow of similar work with animal pathogens which took place at the same time.

This volume will present several groundbreaking papers on fire blight from three scientists who participated in this significant research in the late 19th century – Thomas Jonathan Burrill, Joseph Charles Arthur, and Merton Benway Waite. The work of these three men, from the late 1870s to the 1890s, respectively, represented a period of amazing discovery when the basic etiology of fire blight was proposed and elucidated. The reason this etiology could be considered amazing was that it was the first case when a bacterium was documented as the cause of a plant disease and, therefore, it established the foundations of phytobacteriology in the United States and, by extension, around the world. The efforts of Burrill, Arthur, and Waite also deserve more attention in the context of the development of general microbiology as well.

Burrill, Arthur, and Waite were not, however, the first researchers to turn their attention to the problem of fire blight. Prior to the work of these three, fire blight had a long and destructive history in North America. It had drawn the attention and dismay of many scientists, both professional and amateur, as well as scientific cultivators and progressive horticulturists who suffered under the seemingly inscrutable and capricious nature of the disease.

Indigenous to North America, fire blight had long coexisted in an ecological balance with native, rosaceous host plants such as hawthorn, mountain ash, serviceberry, and wild *Malus* species. But this quiescent situation began to change when European pomaceous fruit cultivars were brought to North America beginning in the 17th century.[4] The first pear, apple, and quince introduced to colonial America probably escaped the ravages of fire blight because scattered plantings and cultivation practices did little to promote vigorous growth while propagation mostly as seedlings limited susceptibility.[5] But as cultural practices changed in the late 18th century, with more widespread planting of fruit trees accompanying the increased focus on the soil and fertilizers and greater use of European cultivars, the disease-causing bacterium spread rapidly.

William Denning of New York made one of the earliest American reports of fire blight in the late 1780s. He observed a "disorder" on apples, pears, and quinces in his Hudson Valley orchards and concluded it was "spreading rapidly" southward. The ailment struck "pear trees and quince trees, to the total destruction of them in a few years." Denning wrote that "this disorder" seemed to be "of a more serious nature than any thing that has ever infested orchards." He discounted the general opinion of farmers that "the trees were blasted by lightning" and believed rather that a worm or borer caused the problem.[6]

In 1817, William Coxe of Burlington, New Jersey, published a description of the disease. Coxe was a successful merchant and typical of the men who joined scientific and agricultural societies and wrote on agricultural topics during this period. He had a great interest in horticulture, although he didn't derive his primary income from it.[7] He was a self-described "scientifick (sic) cultivator" who found his orchards to be "inexhaustible sources of intellectual occupation."[8] Coxe wrote one of the first comprehensive American books on pomology, *A View of the Cultivation of Fruit Trees, and the Management of Orchards and Cider*. In this book he described a disease affecting certain pear cultivars and identified the condition as "fire blight." He wrote "blight, which is sometimes called the fire blight, frequently destroys trees in the fullest apparent vigour and health, in a few hours, turning the leaves suddenly brown, as if they had passed through a hot flame, and causing a morbid matter to exude from the pores of the bark, of a black ferruginous appearance." Coxe concluded that the malady was caused by the sun's rays in combination with humid conditions. He believed that the repeated plantings, or as he termed it, the "long duration of the variety," exhausted stock, and he made the astute observation that overpruning of the pear tree made them susceptible to the blight.[9]

By 1844, the celebrated horticultural writer and clergyman Henry Ward Beecher noted fire blight's movement westward from the Atlantic coast and wrote that already in Indiana and Ohio, the "blight has prevailed to such an extent as to spread dismay among cultivators."[10] One disheartened grower wrote in 1847 that "to see a fine healthy tree, apparently in the condition of the utmost vigor, suddenly – sometimes in a single day – turning black in its branches, and dying as if struck by lightning; this is indeed discouraging to the cultivator."[11] A horticulturist in Illinois bleakly wrote in 1854 that "whilst we justly extol the pear, and assign it the highest place in the list of *exquisite* fruits, our enthusiasm is tempered by the recollection of the . . . maladies which threaten its utter extinction in this country."[12]

The continuing outbreaks of fire blight aroused a growing flurry of attention in the agricultural press, particularly in regards to the debate over the disease's cause. Popular causal theories included lightning, insects, overheating of the pear tree by the sun, and debilitation of stock due to cultivation.[13] The insect theory became the most popular by the 1830s. Many growers naturally made the logical association between the fire blight symptoms and the repeated presence of minute insects, particularly the borer *Scolytus pyri*.[14]

Around 1840, the "frozen-sap" theory began to gain support. This idea asserted that the freezing of unripened wood in autumn and winter manufactured a poison that was transported through the tree by the sap in spring and summer.[15] The frozen-sap theory collided with the insect theory. For example, Henry Ward Beecher supported frozen sap as the culprit. He observed the ravages of the disease in several regions and correlated it to weather patterns and cultivation practices. He, therefore, dismissed the theory that "small, red . . . insects, briskly moving from place to place on the branches" of pear trees, caused fire blight.[16]

Remedies to control fire blight were almost as numerous as the causal theories. Many of these cures reflected "tried and true" agricultural practices gathered over the years or centuries, but some had their origins in human medicine and the theory that health required a balance of bodily humors. For example, some fruit growers combated tree fruit diseases by placing iron around diseased trees, presumably to revive the vital spirits or balance the nutritional needs of the plant.[17] Therapeutics often consisted of liquid concoctions applied to the trees by syringes, brushes, or by washing (painting) the plants with rags or sponges.[18] One of the most famous during the late 18th and early 19th century was the "composition," developed by William Forsyth, the gardener to the king of England. Forsyth's "composition" was a mixture of cow dung, lime, wood ashes, river sand, and urine.[19]

On the other hand, an American horticulturist who believed that insects were responsible for fire blight advised applying chloride of lime to destroy the insect and as "an antidote to the poison" left in the tree.[20] Another insect theorist advocated the use of spirits of turpentine "on and about the diseased part."[21] On the other hand, many frozen-sap theorists advocated cultivation practices such as the use of certain soil types, the selection of naturally early growing and ripening cultivars, and root pruning.[22]

Sulfur was a common catch-all plant disease remedy of the day, and in many cases, a successful one. One grower gave the following sulfur remedy, which he claimed "never failed to revive" the pear tree from the blight:[23]

> I make an incision in tree with an auger, about two inches deep, and in a slanting position, so that the bottom or farther point of the hole shall be inclined towards the roots and generally about two feet above them; the cavity is then filled with sulphur and plugged up perfectly tight: I then cut off the plug smoothly even with the bark. This operation I perform in the month of March or April, or any time when the sap is circulating in the tree.

Many orchardists shared a sense that the disease was contagious, even before scientific evidence of the germ theory. An often-repeated control suggestion to prevent the spread of the disease was to remove and burn diseased branches before the fire blight attacked the entire tree or to remove infected trees altogether.[24] Even with the sound advice, growers felt generally helpless to prevent or retard the injurious force of fire blight in their orchards. "It is so destructive to the pear tree in some sections of the country," one horticulturist lamented in 1840, "that its culture is almost given up."[25]

In the 1850s, Anton DeBary's epochal experiments in Germany demonstrated the germ theory of disease with respect to plants. With scientific finality, he showed that pathogenic fungi could cause disease, in this rusts and smuts of cereals. This revelation meant that fungi became popular suspects as the cause of fire blight by the 1860s and 1870s. In 1863, for example, J. H. Salisbury and C. B. Salisbury of Ohio blamed *Sphaerotheca pyri*, a form of powdery mildew, for the blight.[26] They also claimed that the fungus responsible for the blight in pear was also the cause of blight in apple and quince trees.

Circumstantial credence was given to the fungal theory when several experiments demonstrated the transmissibility of the blight from infected to healthy trees. Reuben Ragan did inoculation tests with pear blight in 1845, and, four years later, Herman Wendell of New York reported that he had "discovered blight to be either contagious or epidemic."[27] In 1868, E. S. Hull, the Illinois State Horticulturist, also transmitted the infection between trees and stated that "fire-blight . . . is induced by an extremely minute fungus, seen only with the aid of a powerful microscope."[28] It is possible that Hull was in fact observing *Erwinia amylovora* but could not comprehend what he was seeing.

It is not unlikely that someone in the 1860s who observed bacteria under a microscope might misinterpret what they were seeing. The study of bacteria was a relatively new field of endeavor for scientists at the time. Hull's observations of fire blight were made barely a decade after Pasteur's groundbreaking work on bacteria and fermentation. In the 1850s, Louis Pasteur had come out of chemistry to demonstrate the active role played by living microorganisms in the fermentation process. His goal had been to persuade biologists that the theory of spontaneous generation was false, and he largely succeeded.

It was not until 1876, however, that Robert Koch in Germany showed for the first time the causal role of bacteria in disease, in his case anthrax in cattle. Unlike Pasteur, Koch came to microbiology via medicine and was more interested in pursuing research for practical reasons of improving public health, rather than to prove or disprove basic theories such as spontaneous generation. He was not the first researcher to make correlations between anthrax and the presence of bacterium in the blood of diseased animals, but he experimentally demonstrated the causal link.

Koch sought out the famous botanist Ferdinand Cohn to confirm his findings on anthrax, and his groundbreaking paper was published in the journal *Contributions to Plant Biology*. This certainly does not imply that botanists were more advanced in bacteriological studies than the researchers on animal/human diseases. Rather, it was simply that Cohn was a well-respected cell morphologist and had done very forward work identifying and classifying bacteria, and he was a scientific figure of great repute in Germany.

Koch made additional major contributions to experimental microbiology with his improvements to pure culture techniques. Since Pasteur's day, scientists had grown bacteria in laboratory flasks filled with some sort of nutrient broth and then transferred the cultures to new flasks many times in an attempt to eventually arrive at pure cultures. It was an unwieldy procedure and invited contamination. In 1881, Koch announced the development of solid nutrient media, at first using slices of boiled potato, but soon moving to gelatin. The solid media technique was further refined in two ways; the first was the adoption of the poured plate method using the Petri dish, and the second was when Fannie Hesse, the wife of one of Koch's co-workers, suggested agar as the solidifying component of culture medium.[29]

During the same period, research on bacterial diseases of animals was going on in the United States, no doubt inspired, at least in part, by Koch's highly celebrated work on anthrax. In the late 1870s, H. J. Detmers of the United States Department of Agriculture was one of several scientists searching for a bacterium that caused hog cholera. Burrill witnessed many of Detmers's experiments and took inspiration from them.[30]

When Thomas J. Burrill began his study of fire blight in the late 1870s, bacteriological work was still in its infancy, particularly in terms of plant diseases. Mitscherlich, a German chemist, had reported to the Imperial Academy of Science in Berlin as early as 1850 that he observed "active liquid" degrading potato cell walls. He blamed this action on *vibrios*, one of the six named genera of bacteria.[31] In 1869, F. M. Dranert studied sugarcane in Bahia, Brazil, commenting that diseased cane produced "yellow material" which "appears as a micrococcus."[32] M. S. Woronin in Russia observed bacteria in root nodules of lupine in 1866, and E. Prillieux in France found bacteria in the cavities of seeds in "rose-red" disease of wheat in 1879. Most of this work, with the possible exception of Dranert's, was inferior to Burrill's fire blight studies because no inoculations were performed.[33] Still, these examples showed that when Burrill turned to fire blight research in the late 1870s, he was not alone in his interest in bacteria.

In addition to a growing awareness of bacteria to researchers in the biological sciences and, of course, the destructiveness of a disease like fire blight which called attention to the urgent need for a better understanding of its cause and control, one last element remained to conspire to create the startling breakthroughs in phytobacteriology in the late 19th century. This final piece of the puzzle was the existence of institutions that could support such extensive work. The story of fire blight also serves as a valuable historical model for studying institutions of science and plant pathology because the three scientists who appear in this volume – Burrill, Arthur, and Waite – represent the three institutions that allowed the science of plant pathology to thrive in the late 19th century in the United States. Burrill worked at a land-grant college, Arthur at a state agricultural experiment station, and Waite at the U.S. Department of Agriculture.

All three of these institutions were founded on the promise of delivering practical support and relief to the farmers of the United States, who still constituted the bulk of the population. These institutions were publicly funded and subject to the pressures of politics. Therefore, their ultimate goal remained serving the nation's farmers by whatever means necessary. Certainly that required engaging in scientific research, but such research always had as its ultimate rationale to solve agricultural problems, rather than for the express purpose of delving into biology for its own sake. During the late 19th century, plant pathology was a new science, and it was overwhelmingly an applied science. Burrill, Arthur, and Waite were drawn to areas of research that were useful in some way to the farmers served by their institutions. Fire blight was a good example.

LITERATURE CITED

1. van der Zwet and Keil 1979; Campbell 1979, 62-69.
2. Vanneste 2000, 1-6.
3. Paulin 2000, 107.
4. Fisher and Upshall 1976, 306.
5. Campbell 1979, 64-65.
6. Denning 1794, 219-222.
7. True 1937, 4-5; Hedrick 1950, 210-211.
8. Coxe 1817, 1.
9. Coxe 1817, 174-176.
10. Beecher 1844, 56.
11. "The Blight in the pear tree" 1847, 59.
12. Overman 1855, 447.
13. For lightning, see "Blight in pear trees, etc." 1829, 57. For insects, see Lowell 1826, 17-18; Buel 1827, 108; Belden 1837, 461; Kenrick 1841, 173-174. For overheating, see Castor 1839, 8. For cultivation, see Kenrick 1841, 174; Beecher 1844, 444.
14. Lowell 1826, 17.
15. Lazell 1840, 415; Beecher 1844, 447-455; Downing 1872, 646-650.
16. Beecher 1844, 452.
17. C. D. 1844, 275.
18. McMahon 1806, 40; Thacher 1822, 181-182.
19. Forsyth 1802, 229.
20. Buel 1832, 54.
21. "Cure for blight of pear trees" 1836-1837, 395.
22. Lazell 1840, 415; Beecher 1844, 455.
23. Waters 1839, 37.
24. Lowell 1826; Kenrick 1841, 173-174; Belden 1837, 461; Elliot 1844 255.
25. Lazell 1840, 415.
26. Salisbury and Salisbury 1864, 452.
27. G. S. B. [Gookins, S. B.] 1846, 254-255; Wendell 1850, 447.
28. Hull 1869, 35; Baker 1971, 612.

29. Reed 1943, 162-163; Coleman 1977, 165-166; Brock 1988, 101-103; Magner 1994, 282-289.
30. Burrill 1879, 80; Burrill 1882, 119-120.
31. Burrill 1881e, 163; Kennedy, Widkin, and Baker 1979, 2.
32. Baker 1971, 612.
33. Baker 1971, 617-618; Kennedy, Widkin, and Baker 1979, 3.

Thomas Jonathan Burrill

Thomas Jonathan Burrill was born in Pittsfield, Massachusetts, in April 1839, but his family moved to Illinois when he was a young boy.[1] He entered Illinois State Normal School in 1862. At the time, Illinois State Normal was the principle institution of higher learning in the state. His professors in the natural sciences were very qualified and included Dr. Benjamin D. Walsh, state entomologist; Dr. George W. Vasey, president of the Illinois State Natural History Society and later a USDA botanist; Professor Jonathan Baldwin Turner, one of the leaders in the national land-grant college movement; and Dr. Joseph A. Sewall, botanist and curator of the Museum of the State Natural History Society. Sewell had been a student of Asa Gray and Louis Agassiz at Harvard, and he stimulated Burrill's love of natural history. In 1865, he graduated with a degree in natural history. Burrill became superintendent of public schools in Urbana, Illinois, until 1867 when he joined the first expedition of Major John Wesley Powell, a fellow Illinoisan, to the Rocky Mountains of Colorado as a botanist.[2] In April 1868, Burrill secured a post at the new Illinois Industrial University, renamed the University of Illinois in 1886, the state's land-grant institution. Hired to teach algebra, within seven months he was professor of natural history and botany, a job more suited to his training and interests. Six months later he was professor of botany and horticulture.[3]

At the Illinois Industrial University, Burrill became one of the first botanists in America to introduce information about plant diseases into his courses:[4]

> A new field of labor is also opening before the student of botany, that of vegetable diseases. Perhaps nothing pertaining to plants is so little understood; but the importance of the study and the increased facilities of late years for microscopic observations will undoubtedly call more attention to the subject.

By 1874, Burrill had initiated the first formal teaching of plant pathology in the United States. His courses on general horticulture, pomology and forestry, floriculture, botany, vegetable physiology, and microscopy all included sections on injurious fungi.[5] He quickly forged a reputation among a very small group of botanists in America who had an "expertise" in plant diseases. One interesting, though possibly apocryphal, story that exemplified his rising stature came from a train trip with Asa Gray and a group of ten or twelve botanists heading for Dubuque, Iowa, in 1872 for a meeting of the American Association for the Advancement of Science. While the train stopped for a wreck to be cleared on the track ahead, Gray stepped off and entered a nearby field. He gathered some smutted corn, returned to the train, and asked the gathered botanists, "Who can tell me what this is? Burrill, keep your mouth shut."[6]

It was not fungi, however, but Burrill's research on the bacterial origins of fire blight that made him one of the most celebrated phytopathological figures of his century. He expressed an interest in fire blight as early as 1873; however, his first major paper on the disease was presented to the Board of Trustees of Illinois Industrial University in September 1876. In this paper, Burrill wrote that he had observed "oscillating corpuscles" in the diseased tissue under a microscope.[7] He did not, however, identify these microscopic corpuscles as bacteria.

He presented his second paper on fire blight to the Illinois State Horticultural Society in December 1877. Again, Burrill reported that the "cambium of the blighted branch . . . is filled with very minute moving particles, very similar to those known as spermatia in fungi."[8] He would later blame his misidentification of fungi on "the want at the time of a proper glass."[9]

Burrill presented his third paper on fire blight to the Illinois State Horticultural Society in December 1878. This is the first of the essential Burrill papers presented in this book. Actually, it is more of a casual discussion rather than a formal presentation, but it is one of the most significant reports in the history of plant disease studies. In it, Burrill tentatively identified the "oscillating corpuscles" as bacteria and even more tentatively suggested that they might be the cause of fire blight. He wrote:[10]

> Does it not seem plausible that they (the particles) cause the subsequently apparent change? It does to me, but this is the extent of my faith; we should not say the conclusion is reached and the cause of the difficulty definitely ascertained. So far as I know, the idea is an entirely new one – that bacteria cause disease in plants – though abundantly proved in the case of animals.

In 1880, there was a severe outbreak of fire blight in Illinois, and in the summer of that year Burrill responded. He carried out an important series of experiments to determine whether a bacterium actually caused fire blight or whether the bacteria were just a by-product of the disease. He inoculated 69 trees on the University campus – 36 pear, 29 apple, and 4 quince.[11] Burrill's inoculum was either exudate or sections of bark from infected shoots. He collected the exudate in the morning from new lesions, diluted it with water, and examined it microscopically to make sure it was free of fungi. Burrill also paid particular attention to the leaves, applying the "virus," as he referred to the exudate, externally in order to determine whether transmission was external or internal. At that time, the term "virus" was used to refer to a poisonous chemical or substance and not to a specific type of pathogen. No external application of the blight exudate resulted in disease transmission, clearly indicating that introduction of the "virus" into the plant was necessary to produce the disease.[12]

The infection experiments were successful enough for Burrill to state that the "introduction of the virus introduced the cause of the disease, and the potency of the virus was quite positively due to the living bacteria."[13] He referred to Pasteur's concept that this microscopic organism might be a common element in many diseases, allowing that if this were so, "it would not necessarily . . . invalidate its agency in producing this disease of the pear; but it might render less hopeful the discovery of remedial treatment." He also showed definitively that the blight of pears, apples, and quince were all interrelated.[14]

Burrill presented the results of his important 1880 inoculation studies in three speeches between August and December, which were all published the next year. First, he presented "Anthrax of Fruit Trees; or The So-Called Fire

Blight of Pear, and Twig Blight of Apple, Trees" to the AAAS in Boston.[15] Also in August, he reported his findings to Regent Theodore Draper of Illinois Industrial College in "Blight of Pear and Apple Trees."[16] In December, he presented a paper that was essentially the same as the AAAS paper to the Illinois State Horticultural Society.[17]

Burrill described the fire blight bacterium in 1880, but it was two more years before he named it. Naming the pathogen was not his priority at the time. In 1881, in a discussion of the history of bacteriology, he wrote that "present information leads me to think the organism of the blight is specifically distinct . . . but this cannot now be definitely decided, neither is it important that we should hasten the verdict. The life history of the creature, its mode of action, our methods of preventing its ravages, are much more important than the name by which it is known."[18] Still, in 1883, Burrill finally named his bacterium *Micrococcus amylovorus.*[19] Unfortunately, the name was based on the mistaken concept that the bacterium was a starch destroyer (*amylo* + *vorus* = starch eater). The species name would stick, however, with modifications over the years. In 1889, it was changed to the widely accepted *Bacillus amylovorus* (Burr.) Trevisan. It went through subsequent less accepted names, first in 1897 to *Bacterium amylovorus* (Burr.) Chester and then briefly to *Bacterium amylovorum* (Burr.) Serbinoff. In 1920, bacteriologists finally settled on *Erwinia amylovora* (Burr.) Winslow et al.[20]

In 1880 and 1881, in addition to naming the microbe, Burrill performed extensive inoculation tests, but he never used pure cultures in his fire blight experiments. He did not attempt to satisfy Koch's postulates by isolating the bacterium, inoculating with the isolate, and then re-isolating the same organism. Thornberry claims that pure culture techniques had not been invented when Burrill did the bulk of his fire blight work.[21] Baker maintains, however, that William Farlow had used pure culture techniques at Harvard prior to Burrill's work on fire blight. Furthermore, Baker asserts that Burrill was aware of the methodology and was always concerned about the infirmity of his results.[22] Burrill could read German and, given his excellent historical overview of the field of microbiology, it is likely he was aware of Koch's definitive work on anthrax. But still, there was an overlap between Koch's breakthroughs and Burrill's fire blight research so it is not incredible to believe Burrill did not incorporate this new and unfamiliar technique in research that he had already undertaken. The answer, however, as to why Burrill did not go back and perform pure culture experiments with the fire blight bacteria remains unclear.

There is evidence that Burrill simply considered the level of proof he provided to identify *Micrococcus amylovorus* as the causal agent of fire blight to be sufficient. He may have done little follow-up work because he interpreted the absence of dissent about his theories as approval. "There seems to have been no attempt," Burrill wrote in 1883, "to disprove the conclusions as published in . . . 1880, nor has there been any evidence . . . that anything besides bacteria does this deadly work in the tissues of our pear trees." [23] Given what he considered to be the overwhelming support of silence, he felt satisfied in moving on to other areas of research and other university duties.

In his research on fire blight, Burrill did not just seek the causal agent. His interest in the disease was literally more than just academic because, as a land grant researcher with a mandate to further public good, he had to pursue aspects of the disease that were more practical and would perhaps point to a solution to the problem. One of these areas of inquiry that was of interest to orchardists was the point of infection. This topic was a matter of confusion and debate. Burrill proposed that the microorganism had to enter the tree through wounds or punctures.[24] He did not consider flowers as a site of infection until it was suggested by George P. Peffer, a grower from Wisconsin.[25] Although Burrill readily accepted flower infection as a possibility after that, the viability of blossom infection remained debatable.

Of course, the most pressing issue for orchardists was disease control. Burrill had to address the issues of preventatives or cures for the disease, but he found he had little to say. Indeed, there was nothing he could recommend beyond some traditional methods long used by growers. He wrote: [26]

> No claim is here made to the discovery of prophylactic or remedial methods which shall change fruit-growing from failure to success. . . . While something is to be advanced in the way of improvement in the means usually adopted to save our pear trees, the information already given may

> be of more service in saving us from crude hypotheses and useless labor, fighting in the dark an unknown foe.

He dismissed the popular notion that mutilating the trees, such as root pruning, served any purpose in combating blight other than to remove the disease by killing the tree. He did warn against the careless pruner's knife serving as a means of creating a wound for possible infection or transferring the infection from tree to tree. He recommended washes of lye, linseed oil, lime and sulfur solutions, and even carbolic acid, not because he had experimental proof that they could prevent or cure fire blight, but because they "do no injury, and may be beneficial." [27]

Burrill's experimental work on fire blight was the pinnacle of his career in scientific research. However, fire blight was not the end of Burrill's bacteriological work. He attributed a variety of plant diseases to bacterial origins. Burrill pointed to bacteria as the cause for diseases of Lombardy poplar, butternut, and aspen, and even suggested erroneously that bacteria were responsible for the "poison" of poison ivy.[28] Burrill had also surmised that a bacterium was the cause of peach yellows, but in 1883 he recanted earlier statements and insisted that "it is impossible for me to have an opinion as to whether bacteria have, or do not have, anything to do with the disease called yellows." [29] He investigated bacterial origins of "corn blight" in 1885.[30] He did use pure culture techniques when he studied broom-corn and sorghum diseases in 1886 and 1887.[31]

Burrill spent his entire career at the University of Illinois, although with time he moved more into university administration and away from research. Still, his body of writing and teaching did much to further plant pathology in the United States and to establish the field of phytobacteriology. He wrote two important general works on the subject: *Bacteria and Their Effects* (1881) and *Bacteria: An Account of Their Nature and Effects, Together With a Systematic Description of the Species* (1882).[32] He taught and molded a generation of students, particularly in the field of microbiology. A bacteriology laboratory established at the University of Illinois in 1892 "offered the best facilities extant at the time in any American University (exclusive of a few medical colleges)." [33] Burrill's students and experiment station assistants included many who would have significant influence on botany and plant pathology, such as George P. Clinton of the Connecticut Agricultural Experiment Station, Benjamin M. Duggar of the University of Wisconsin, Arthur B. Seymour of Harvard, and Merton B. Waite of the United States Department of Agriculture.[34]

Burrill retired from the University of Illinois in 1912 and died four years later. He had a long career of varied research areas including extensive work with many diseases of fungal origin as well as prodigious efforts in taxonomic mycology. However, it was Burrill's work on fire blight that endured and marked him in history, and he lived to see this pioneering work widely accepted. But like many pioneers, he is not just noted for the trails he blazed, but for the paths he left unbroken. Burrill left phytobacterial studies well begun but in an imperfect state.

Burrill produced many papers on fire blight that merited reprinting in this book. However, choices had to be made and the two presented here are of special worth. The first may seem an odd selection at first. It is a paper dealing with fire blight which he presented before the Illinois State Horticultural Society in December 1878. In point of fact, it is not a paper at all but more of a record of a discussion held during the meeting. Still, it should rank as one of the most significant reports in the history of plant disease studies as it is the first time Burrill, or anyone for that matter, makes the claim for a bacterial cause of plant disease based on any sort of evidence. But the nature of plant pathology at the time can be judged by noting that this groundbreaking research is met with only limited interest in the part of Society members and is sandwiched between much more spirited discussions of the most productive varieties of fruit trees.

The second Burrill paper reprinted in this book is "Blight of Pear and Apple Trees" which originally was published in the Illinois Industrial College report for 1881. Although "Anthrax of Fruit Trees," which he presented before the AAAS, could be considered the first official presentation of Burrill's research, and is certainly a more famous and widely cited paper, "Blight of Pear and Apple Trees" is in fact a more illuminating discussion of the work. In it, Burrill gives not only a good historical overview of the disease but also outlines the very limited state of plant bacteriology as he knew it from primarily European sources. In addition, he discusses his experiments that led to the discovery of the causal organism in much more detail than in "Anthrax," and he provides a more detailed description of the bacterium.

More important than just the content that is expanded when compared to "Anthrax," however, is Burrill's obvious interest in fire blight in an agricultural, not just biological, context. He clearly demonstrates the overwhelmingly practical demands on plant disease studies as practiced in a land grant college of the day. Plus, "Anthrax of Fruit Trees" has been quoted so often in the past that a fresh view of this groundbreaking research is highly desirable. Therefore, we believe the two Burrill papers presented here are illuminating examples of his work on fire blight and an excellent snapshot of plant pathology and phytobacteriology at the very dawn of the science.

LITERATURE CITED

1. Barrett 1918, 1; Glawe 1992, 17-18.
2. Barrett 1918, 1; Thornberry, H. H. 1964. Thomas Jonathan Burrill's contribution to the history of microbiology and plant pathology. In Thornberry Papers, 1929-1994, 24-25; Solberg 1968, 125.
3. Thornberry, supra note 2, 25-26.
4. Burrill 1869, 327.
5. Thornberry, supra note 2, 28.
6. Hottes 1940, 6-7.
7. Thornberry, supra note 2, 34.
8. Burrill 1878, 114.
9. Burrill 1881d, 73
10. Burrill 1879, 80.
11. Burrill 1881b, 527-530.
12. Burrill 1881a, 583-597.
13. Burrill 1881d, 69.
14. Burrill 1881e, 157-167.
15. Burrill 1881a, 588-596.
16. Burrill 1881d, 62-84.
17. Burrill 1881e, 157-167.
18. Burrill 1881d, 79.
19. Burrill 1883a, 319-32
20. van der Zwet and Keil 1979, 37.
21. Thornberry, supra note 2, 38.
22. Baker 1971, 621.
23. Burrill 1883b, 46.
24. Burrill 1881d, 80.
25. Peffer 1882, 191-202; Burrill 1883b, 49.
26. Burrill 1881e, 164.
27. Ibid., 165.
28. Burrill 1881c, 271-273; Burrill 1882, 134.
29. Burrill 1883b, 48.
30. Burrill 1889, 19-25; Moore 1894, 368-369.
31. Burrill 1887, 30-36.

32. Thornberry, supra note 2, 38.

33. Burrill, T. J. to E. J. Townsend, February 23, 1907. In Thornberry Papers 1929–1969.

34. Thornberry, supra note 2, 30.

Fig. 1. T. J. Burrill, ca. 1880. (Courtesy of the Iowa State University Library, Department of Special Collections.)

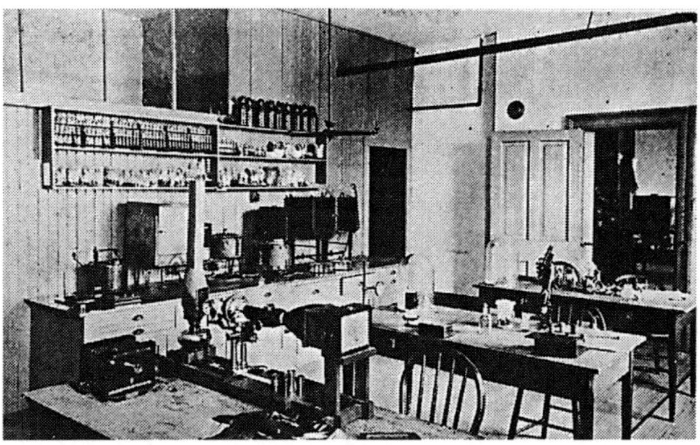

Fig. 2. T. J. Burrill's bacteriological laboratory in University Hall basement, University of Illinois, 1880s. (Courtesy of University Archives, University of Illinois.)

Discussion on the Reports from the Transactions of the Illinois State Horticultural Society, 1879.*

The President asked Mr. Hammond if he had observed any rust on the Snyder blackberry or others.

MR. HAMMOND. – I have seen the orange rust on the Kittatinny, but not on the wild blackberries.

THE PRESIDENT. – I have not seen the rust upon the Snyder, while the Lawton growing side by side with it was almost killed out by rust; the wild blackberries, also, have rusted badly. The statement in one of the papers read that seedling peach-trees are hardier than budded ones is substantially an error, and should be noticed. Hardiness or tenderness may be propagated by budding or grafting as well as through seeds. As a general rule, we know that budded fruit is superior to seedlings, and it is also true that the trees may be equally hardy, or even more so, if pains be taken to propagate from hardy varieties; but so long as we raise Early and Late Crawfords extensively, so long will the idea prevail that seedlings are the hardiest.

A Voice. – Do you know anything about the Barnard and Slocum?

THE PRESIDENT. – The Barnard tree is very hardy, the fruit of good size, deep, dark colored on the outside, yellow flesh, and firm, of an aromatic flavor. It ripens about the same time as Early York. The Slocum ripens about the same time; but is quite subject to the attacks of the curculio, and I do not consider it a desirable variety.

DR. LONG. – Do you know anything about the Alexander?

PRESIDENT. – I have not fruited it myself, but I hear it spoken of very highly. Doing well in Texas.

DR. LONG. – Do you know anything about Haines' Early?

PRESIDENT. – It is not as early as Early York.

PROF. THOMAS. – I have found some seedlings in Southern Illinois that are very early and fine.

PRESIDENT. – By planting seeds of the early kinds of budded fruit they will generally produce early varieties. The new kinds we originating now may be hardy and free from the yellows. Alexander and Amsden are thought to be the same.

*Originally published: 1879. *Transactions of the Illinois State Horticultural Society for 1878.* Chicago: Illinois Horticultural Society. 76-81.

DR. LONG. – I went into the business of growing fruits in early times; I began in 1832, selecting from the best fruits I could find and people were astonished at my results. We had no insect-enemies at that time to trouble us, none even on the wild fruits. I began in 1858 with pears, collecting some sixty or seventy varieties, but have reserved only six or seven; they have all blighted more or less. They seem to blight more in seasons when they make a vigorous growth, and especially a heavy thunder storms. I do not think the pear needs much cultivation, and but little manure and less pruning. I have seen Flemish Beauty growing in ground that had never been broken up, doing well; also White Doyenne. I think we can still raise good peaches; but I have not yet found the apple to fill the bill.

The Doctor explained the way in which he propagated apple-tree at an early day; which was to take young wild crab-apple trees, bury their bodies in a shallow trench, letting the tops stick out, bore holes along bodies, and insert cions (sic). Roots would form along the trees, the cions (sic) grow, and thus, by taking up and cutting apart, independent grafted trees were obtained.

A Voice. – Which is the most productive peach?

DR. LONG. – I don't know; Crawford is good. Ward's Late (free) comes in after Old-Mixon cling.

PRESIDENT. – Venus is a small peach, of fine flavor, and the tree is very productive.

MR. McWHORTER. – I would not like to have it go on record that a worm causes what is called fire-blight, as stated in one of the sub-reports.

PROF. THOMAS. – I don't believe any one knows the cause of twig-blight or fire-blight. I have examined carefully with a glass, and cannot find any insect; the twigs look as though they had been struck by lightning. It may be caused by electricity. The spur-blight is doubtless caused by an insect.

PROF. TICE. – Mr. President, I would not think of participating in this discussion if I did not consider that meteorological causes had something to do with the disease known as blight. I have observed that blight makes its appearance suddenly, generally in June or early July, and within a very few days after violent thunder storms. I hence have an hypothesis that electric explosions have in some way an intimate connection with it, and may possibly stand to it in the relation of cause. The nature of electric discharges is not so well understood as it should be. It is the general notion that the discharge only occurs at a point from which the discharge breaks or is received. Now, such is not the fact. Every spear of grass, the twig of every tree, and especially the needle-leaves of pines, for a mile or more, discharge simultaneously with the visible discharge. It has been ascertained by repeated experiments that a wire coiled at one end around a steel wire will magnetize the steel, even if a discharge occurs fully a mile and more distant. This is caused by induction, which generates a current in the wire, and consequently in the coil.

Good housewives know that thunder sours the milk. It is not by any visible discharge in the milk-pans, but again by induction; the positive overhanging cloud evokes by induction a negative discharge from the earth beneath, which is done in the following manner: The cloud separates the opposite polarities, which are in equilibrium in the earth, driving the similar polarity, the positive, into the earth, and attracting the dissimilar polarity, the negative, in high tension on the surface of the earth beneath. A portion of the negative is lost by neutralizing the positive air in contact with it, but a greater portion by instant conduction to the point of discharge. The positive tension on the cloud being relieved by the discharge, the positive that had been driven into the earth surges back and produces what is called a return stroke. This return stroke decomposes the mllk (sic), and consequently it sours.

The blight, when the trees are in bloom, is effected (sic) in the same way, the nectar of the pistil is decomposed both by the discharge from its summit and by the return stroke whenever lightning strikes in the vicinity; it therefore sours, and fructification cannot take place. I think there is some thing similar takes place to produce the apple and pear-blight, especially the latter, for it is well known that fermentation takes place in the sap; it has turned sour, and scalds the tree as it descends. I present this not as a theory, but as an hypothesis to be confirmed or disproven by observation. I would therefore suggest that the members make observations upon this point next season and report the facts, and then we will be at no loss to discover the theory that explains them.

MR. McWHORTER. – Will Prof. Tice explain why some varieties are more subject to blight than others, upon that hypothesis.

PROF. TICE. – I suppose it is upon the same principle that some persons are delicately organized and so nervous that they cannot endure living in houses constructed wholly or in part of iron; the electric currents surging constantly to and fro in the iron cause severe headaches and general physical prostration, while more vigorous persons are unaffected.

PROF. BURRILL. – The so-called fire-blight of the pear and common twig-blight of the apple are almost surely identical in nature and as surely due to the same cause. Other agencies may, in special cases, produce the death of limbs of these trees, but the wide-spread and well-marked maladies of which we speak are not caused by insects or any other living things large enough to be discovered by the aid of a hand magnifier, much less by the naked eye. In my observations the Siberian crabs and some varieties of Russian apples are, in regard to this disease, intermediate between the pear and the common varieties of *Pyrus Malus* (sic). Indeed, some crabs are almost as badly injured as the pear.

It is curious how we see through our own spectacles. The entomologists look for insects' work, the physicist finds the cause in electricity, the botanist discovers physiological changes, or attributes the effect the action of fungi, etc. You know, perhaps, through what kind of glasses I have been gazing and anticipate my speech. There are evident enough numerous species of fungi on the limbs after death has taken place, but no one can point out how any one of these is related to the disease itself, what particular one, if any, is at the bottom of the trouble, and in what manner it lives and operates to cause the mischief. Some other disease of the pear, apple, etc., can be directly and certainly traced to a special parasitic fungus, but this has not been done for this blight.

This allusion to sour milk, however, ought not to go by without notice. Perhaps the phenomena in the one may help us to explain the other. It so happens that my attention has been recently called to the changes which milk undergoes in becoming sour, and in putrefying, and I have had a chance to verify, by numerous experiments, the theory that these phenomena are caused by microscopic organisms, germinating, living and multiplying in the fluid. In milk kept in a warm place the minute organisms can be found in considerable number before the acidity can be detected, except by chemical tests, and their number increases with the progress of the change, becoming enormously great. A drop of the fluid thus populated added to fresh milk produced the result very much sooner than it would have occurred without it. Milk boiled and hermetically sealed keeps forever in spite of electricity, etc. So, too, a little carbolic acid, salicylic acid, and other compounds which kill these microscopic creatures, added to milk prevents fermentation, and the otherwise unstable fluid remains as it was in the beginning.

Now, returning to the disease we are discussing, a startling correspondence with the foregoing may be observed. If we remove the bark of a newly-affected limb, and place a little of the mucilaginous fluid from the browned tissues under our microscope, the field is seen to be alive with moving atoms known in a general way as bacteria. Sometimes a thick, brownish fluid oozes from the bark of dying limbs and spreads over the outside or falls in drops. This is apparently made up of the living things, myriads of them to be seen at once. A particle of this viscous fluid introduced upon the point of a knife into the bark of a healthy tree is in many cases followed by blight of the part, but with me not in every instance. The result, if such it be, is not so uniform as the infection of warm milk; for this, however, there are very possibly other explanations. It is not improbable that certain trees in certain conditions are able to withstand the destroying influences of lowly-organized plant enemies. Plants and parasites are forms of vegetable life; in the struggle for existence one or the other is subdued and the vanquished perishes. If we look once more to the affected branch we find the disease spreads more or less rapidly from the point of origin, and upon examination the moving microscopic things are discovered in advance of the discolored portions of the tissues, but not very far ahead – an inch, perhaps. Does it not seem plausible that they cause the subsequently-apparent change? It does to me, but this is the extent of my own faith; we should not say the conclusion is reached and the cause of the difficulty definitely ascertained. So far as I know the idea is an entirely new one – that bacteria cause disease in plants – though abundantly proved in the case of animals.

Dr. H. J. Detmers, of the United States Commission, thinks he has established the fact that the so-called hog-cholera is due to such organisms, and I have repeatedly seen them swarm in the blood, bile, etc., taken from the diseased animals just after death occurring in the usual way or from violence. The so-called germ-theory of disease in animals, especially in infectious and contagious diseases, is rapidly gaining support and credence. It is not impossible that we are now making a beginning of the application of the theory to the diseases of plants which have heretofore been mysterious and inscrutable.

PROF. TURNER. – I have had quite a struggle with this blight, and as far as accounting for it is concerned have always come out upside down. There are three forms of blight: twig-blight, which is confined to the young tender shoots; the spur-blight, which only kills leaves and fruit on the fruit-spurs; and the fire-blight or pear-tree blight. I have never found this fire-blight showing itself where there was not a black patch the previous year on the affected branch below where the blight showed itself. We often do things year after year to obtain certain results, without knowing why the results follow. I have arrested the fire-blight – as I have before stated in these meetings – by shaving off the bark from those dead patches and painting them over with turpentine and lamp-black. I found that this disease is exceedingly contagious; for if I used my knife to prune a healthy tree after having used it in shaving the diseased one I communicated the disease to that tree; so I soon learned to dip my knife in spirits of turpentine and wipe it off before using again. I didn't exactly know what I was doing, but Prof. Burrill has just now told me: the spirits of turpentine killed the living things which were present.

I have found that coal-oil applied to the surface of the throat and rubbed in will arrest the croup every time. I have also used sulphur for the same disease, dusting it inside the throat. Now, if the lining of skin is filled with living things, as Prof. Burrill tells us is the case in fire-blight, the coal-oil killed them. The longer I live the more I become convinced that nearly all forms of disease result from some new forms of life – usually microscopic. The fire-blight maybe first caused by electricity.

PROF. BURRILL. – I have seen Kittatinny and Snyder blackberries growing in the same row, and the Kittatinny all killed by the rust, while the Snyders were not injured. I know a man who sold eight hundred dollars' worth of fruit of the Snyder from one and seven-eighths acres the past season.

A Voice. – Will Mr. Minkler name the best apple for market?

MR. MINKLER. – The old fossils don't have much to say. I wish ask Mr. Hammond if the Ben Davis is not to-day the most popular apple in St. Louis? We must raise something we can sell. People will give no more for Grimes' Golden or Jonathan than for some of the poorer varieties that are showy. We can grow the Ben Davis at one-half the cost of most sorts. The tree is a very rapid grower, and I am planting them for live fence-posts. Apples have been so cheap this year, selling at thirty, forty and sixty cents per barrel, and owing to the low price the packing has been poorly done, and apples are not keeping well.

MR. HAMMOND. – The Ben Davis is very popular for Southern trade, but Chicago don't want them. When all our orchards of Ben Davis come into bearing the fruit will not be worth more than fifteen cents per bushel. They are not good cooking apples, and only fair for cider.

Blight of Pear and Apple Trees.*

T. J. Burrill

The destructive disease of the pear tree, known as Fire Blight, and the prevalent Twig Blight of the apple tree, as well as the injury to the trunk usually called Sun Scald, in one or both trees, are but varieties of the same malady. The same cause is at the bottom of these, and other forms of the disease; and other trees and plants similarly suffer. The blight of the quince is identical with that of the pear and apple tree. The Yellows of the peach, if not caused by the same agent, is certainly produced by a closely allied parasite. The Lombard poplars are dying all over our country from, first, (at least in many cases,) the work of an insect, followed by the microscopic living things which are here charged with the destruction of so many fruit trees. The butternut tree has been considered almost worthless from the liability to a disease of the trunk similar in every respect to the so-called Sun Scald of the apple. The twigs of the aspen poplar blight just as we observe the disease in the shoots of the apple, and from the same or very similar cause. The disease is much more progressive in some seasons of the year than others; but it is not confined to spring, summer, autumn or winter. During warm days of winter certain progress has been observed, though closer watching is required to detect the equally fatal work of the destroyer. Meteorological conditions may have some effect, but thunder showers have little or no influence upon the progress of the disease.

The direct cause of the serious and hitherto mysterious malady is a minute organism belonging to the lowest order of Fungi, (or, as some say, Algæ,) best known as *Bacteria*. Similar organisms are called "Disease Germs," and are supposed to cause deadly maladies in animals and man. That plant diseases are also produced by them had not been intimated until the facts presented in this paper were made known by the writer.[1] A short historical account of the disease and the literature upon it, without the least claim to

[1]See especially "Anthrax of Fruit Trees," in Transactions of the American Association for the Advancement of Science, 1880. The present paper is in part a transcript of this.

*Originally published: 1881. *Illinois Industrial University Board of Trustees Tenth Report for Two Years, ending August, 1880*. Springfield, IL: H. W. Rokker. 62-84.

fullness or completeness, is presented, introductory to a special account of the investigations and experiments by which the true cause of the difficulty was discovered and proved.

HISTORICAL SKETCH

The peculiar disease called Fire Blight of the pear, and the same thing with a somewhat different appearance in the apple, hence known by other names, has been known in the United States over one hundred years. During some periods the injury has been inconsiderable; at others very great. Pear trees have especially suffered, and on account of this disease alone this favorite fruit is scarcely planted at all in many great sections of the country. Practical horticulturists have considered the risk too great to run, except upon the smallest scale without thought of profit. Apple trees are less frequently killed outright, though it often happens, especially when the trunk is the seat of the disease. More often only the young wood of the current year's growth dies. The quince is not less injured than the pear. The peculiar disease of the peach tree, called the Yellows, seems to have been somewhat later observed and described; but it has no doubt existed from the time peaches were first cultivated in America. The first notice gaining publicity was in the Atlantic States, southward from New Jersey. The spread of the disease is more clearly traced in new regions from the contagion of the old, than is the case with that of the pear and apple.

No topic has been more fully discussed in the meetings of horticulturists, and none has a fuller literature in the periodicals devoted to rural affairs. Theory after theory has advanced as to cause and cure, only to be overturned by subsequent observation and experience. Insects were charged with the mischief, and one or more species, actually causing at times death of the limbs, have been described as the true cause or causes of the disastrous disease. These publications becoming well known, satisfied unskilled observers, and hundreds of authoritative statements have barely sufficed to convince many persons to this day of the error. But it is sufficient to say that no competent entomologist now attributes the devastating blight of these trees to the direct work of insects.

Downing argued[2] – and his teachings were widely accepted – that the difficulty originated in the freezing of the sap, and its subsequent corruption in the tissues, hence the name Frozen Sap blight. But the disease occurs when no such conditions are possible, as in the case of seedlings of one summer's growth. Careful observations upon older trees have not supported the conclusions reached by this eminent author. Others believed they had traced the origin of the disease to the absence of essential elements in the soil, and numerous, indeed, have been the proposed remedies based upon this idea. As in other cases, some of the experimental trials seemed for a time to substantiate the conjecture, and forthwith a cure was heralded over the land. The published literature upon this notion alone would fill a volume, but if brought together would very soon furnish its own refutation. Many of the speculations were carefully based upon known facts of vegetable physiology, but most of them have been crude and wide of truth. Again the organic debility of the tree has been considered the cause of its sickness and death. The artificial methods of propagation and treatment were supposed to have wrought injurious constitutional changes, affecting the structure or the vitality of the tree, and death ensued either as direct consequence, or through inability to withstand the ordinary vicissitudes of plant life. It is a hard matter to prove or disprove such a theory as this; the facts, if there are any, are too subtle for exact observation; but we know that seedlings are as liable to blight as nursery varieties, and that the same varieties are subject to the disease in one location and not in another. Unless specific and well grounded facts can be cited, such theories are worthless, and worse than worthless, for, by covering up our ignorance with apparently learned words, we prevent true observation and research by ourselves and by others.

Finally, when all visible or tangible agents were given up, and the clash of theories and speculations showed how utterly inadequate they were to account for the phenomena observed, the idea of destruction by Fungi was advanced. This, too, was a mere speculation. These vegetable parasites were known to cause some diseases of plants. Microscopes came into use, and revealed the minute creature in the very act of parasitical warfare upon their suffering hosts.

---

[2] Fruits and Fruit Trees of America, 2nd ed., p. 646.

To be sure, there was a supposition a long time current that previous to, and necessary to, the attack of these enemies, there must be some constitutional or other weakening of the vitality of the plant; but really little was known of fungous parasitism. The supposed offenders were only seen by expert microscopists, and these were rarely vegetable physiologists. It is not to be wondered at that gross mistakes were made, and that announcements by really scientific investigators proved as unreliable as the hap-hazard (sic) crudities of previous writers. I here give a short sketch of the more important publications upon this portion of our subject. It will be readily seen that, notwithstanding all that had been said about the fungous origin of the blight of the pear, very little was really known about it. So, of course, of the disease in the apple and other trees, as certainly so of, the Yellows of the peach.

In 1863, Dr. Salisbury[3] of Ohio, published an illustrated account of his microscopical investigations, concluding that blight was due to a fungus which he identified as belonging to a well known class of parasites living externally upon the epidermal surfaces of plants. Among these, the European mildew of the grape vine, that of the hop plant, of cherry leaves, of verbenas, etc., are properly classed. Dr. Salisbury named the fungus found by him *Sphærotheca pyri*. Now, it is a fact that such a fungus parasite is not uncommonly found upon the pear, the apple, and other trees, and it certainly is to be credited with its own peculiar injury to the trees; but the fungus was not at the time specifically new to botanists, and has no connection with the deadly blight. Yet this paper is frequently quoted as an important one, and pardonably so, by those not better informed, for the results were apparently based upon direct and skillful investigation. It is the only instance in the whole literature of the subject where claim is made to the identification of the specific parasite causing the disease. Though it has been fully and clearly known for years that this claim was worthless, the attention thus called to the subject has been of much value. The fungus had been previously named, so that there is nothing left except the history of the investigation, to connect the publication with the disease as it is now known.

[3]Transactions Ohio State Agricultural Society, 1863, p. 450.

Dr. Hull,[4] of Illinois, in 1868 and 1870, gave an account of observations and experiments proving, as he supposed, the accuracy of the fungoid theory. One his statements is of especial interest in this paper. He had received through the mail a blighted apple twig, and remarks: "We cut from it several small slices of bark, going deep enough to include a thin slice of wood; with this we inoculated several succulent pear shoots, by tieing in the pieces of bark as in budding. * * * After a lapse of thirty-four days, * * * we found them all blighted." He also adds that the conditions of the weather were not the most favorable for it, though a warm drouth was prevailing, and he believed "a humid condition of the atmosphere is also required." There are other statements by observers that blight is communicable from infected to healthy trees, but there does not seem to have been any carefully conducted experiments upon which such statements are based. In 1868, Dr. Hull thought he had induced disease similar, at least, to fire blight, by grinding in water fibres of fungi found upon decaying roots of living pear trees, and inoculating with the resulting mixture. The direct contagion of the yellows of peaches has been more definitely known. The spread of the disease gradually through an orchard from the point of infection has been frequently observed, if we may trust the accounts given. The first introduction to an uninfected region of country has been more conclusively traced; but here, again, accurate experiments to ascertain the nature and specific action of the contagion have not been made. Theoretically, fungi have been accounted the destroying agents, but with no better evidence than the want of other explanation, and, at most, a mushroon-like (sic) smell of freshly exposed roots and the earth about them. After a series of careful microscopical investigations of diseased specimens of peach limbs, leaves and roots, a thoroughly competent authority,[5] looking for any of the hitherto known parasitic fungi, tells us none such can be found to which the yellows may be attributed.

Returning again to the blight of pear trees, it is worthy of record that in an address before the American Pomological Society, at St. Louis, 1866, Thomas Mehan, of Pennsylvania, gave his reasons for theoretically holding

---

[4]Transactions Illinois State Horticultural Society, 1868, p. 35, 1870, p. 220.

[5]Byron D. Halsted, in Transactions of American Association for the Advancement of Science, 1880.

that the disease is caused by fungi. He has steadfastly adhered to this opinion, and has year by year pointed out, through the columns of the Gardener's Monthly, and elsewhere, the fallacy of other explanations, together with the apparent evidence supporting this. At his instance, Dr. J. Gibbons Hunt, of Philadelphia, examined, by the aid of the best microscopical equipments, blighted limbs of pear sent to him for this purpose. He reported the discovery of a fungus which, according to him, caused the blackening of the branches. In his communication he says:[6] "It attacks the bark and outside of leaves and young fruit first, causing changes in the cells in these locations, resembling much those pigmentary cell-changes which differentiate the negro from the so called white man. The cell contents, normally transparent, are changed into extremely minute pigment granules, which fill the cells and give that characteristic color and smell which mark the disease. Moreover, minute drops of viscid, offensive liquid come out on the surface. * * * From the cambium layer the fungus travels towards the interior of the stem, through the medullary rays chiefly, and here I find those round bodies which, in our hasty ignorance, we often call spores. The ducts which ascend the stem are often obstructed with similar bodies and aggregated pigment granules. This is all I know about the subject. I cannot venture to name the fungus."

I have thus quoted at some length this account, for, according to my observations, it is the nearest approach to the solution of the cause of the difficulty, founded upon direct observation, that had up to this time been made. Still, it seems impossible to admit the correctness of the interpretations here given. The coloration of the diseased parts does not come from pigment granules, or from anything which can be mistaken for them. As for the round bodies which might be called spores, in the interior of the stem, obstructing the ducts, etc., I can only say nothing similar has been found by myself. The "offensive liquid" seems to have escaped careful examination.

In 1876, Thomas Taylor exhibited drawings at the Centennial Exposition, Philadelphia, showing[7] "the effects of the chemical changes which take place in the interior structure of the tree, under the attacks of the fungus

---

[6] Gardeners' Monthly, 1875, p. 245.

[7] Report of Department of Agriculture, Washington, D.C., 1876, p. 75.

to which this disease is due." Unfortunately, I do not remember how nearly these drawings represented what I have observed. They have not been published; neither has the author made any claim of having seen the "fungus to which this disease is due."

In 1877, the writer presented to the Illinois State Horticultural Society the results of microscopical observations, in which account occurs the first published statement of which I have any knowledge, of the minute moving organisms now known to be always present in the portions of trees suffering from the diseases of which we write. I quote:[8] "The cambium of the blighted branch, when trouble first shows itself and for some days thereafter, is filled with very minute, moving particles. * * * Not unfrequently, a thickish, brownish, sticky matter exudes from affected limbs, sometimes so abundant as to run down the surface or drop from the tree. This proves to be identical with that noticed in the cambium, and unquestionably has the same origin. The sticky, half-fluid substance thus exuding is entirely made up of these minute, oscillating particles." In a subsequent discussion, 1878, in the same Society, I am credited with the following:[9] "If we remove the bark of a newly affected limb and place a little of the mucilaginous fluid from the browned tissues under our microscope, the field is seen to be alive with moving atoms, known in a general way *bacteria*. * * * A particle of this viscous fluid, introduced upon the point of a knife into the bark of a healthy tree, is in many cases followed by blight, – but, with me, not in every instance. * * * If we look once more to the affected branch, we find the disease spread more or less rapidly from the point of origin, and upon examination the moving, microscopic things are discovered in advance of the discolored portions of the tissues, but not very far ahead – an inch, perhaps. [They are now known to advance sometimes several feet ahead of the blackened parts.] Does it not seem plausible that they cause the subsequently apparent change? It does to me, but this is the extent of my own faith; we should not say the conclusion is reached, and the cause of the difficulty ascertained. So far as I know, the idea is a new

[8] Transactions State Horticultural Society, 1877, p. 114.

[9] Same series, 1878, p. 80.

one, that *bacteria* cause disease in plants, though abundantly proved in the case of animals."

This meagre sketch of the literature of our subject omits much more than it includes; but it is believed that every item of special bearing upon the essential point of this paper has been noticed. After all the investigations that had been made, and all that had been said – a kind of ferment, in itself – editors were obliged to answer inquiries as follows, quoted from the American Agriculturist:[10] "We are obliged to reply, as we have many times in the past, that the 'cause' of the blight is not known. It is supposed to be due to fungi, – but this has not been demonstrated. * * * Though pear blight is a topic discussed at pomological meetings more than any other, scarcely any positive knowledge has been thus far elicited." This truthfully expresses the knowledge, or rather want of knowledge upon the subject at the time it was written. If what follows may be admitted as an explanation of the specific cause of the blight, certainly some progress has been made. This is the firm conviction of the writer, and this the claim for which he seeks the scrutiny and judgment of those scientifically or practically interested.

As opportunity again presented itself in June, 1880, further investigations were undertaken to ascertain the presence and action of the minute organisms heretofore mentioned as occurring with blight. They were again found in every limb and scarred patch upon trunk in an active condition of disease. Not a single exception was found in hundreds of examinations. More careful study also showed, that so far as could be determined with a microscope furnished with Spencer's first class one-tenth inch objective, affording a magnifying power of one thousand diameters, with the clearest and sharpest definition, the organism was, wherever found, one and the same specific thing. It belongs to a group of very simple plants which, as a class, are as nearly ubiquitous and omnipresent as material things can well be. Wherever organic matter is in any measure undergoing the process of putrefaction or decay, wherever compounds containing sugar or other carbo-hydrates (sic) become sour, wherever ulcerations and pus formations take place in the tissues of animals, *bacteria* of one kind or another are found. On the other hand they are not

---

[10] Vol. 37, p. 356. (1878.)

found in substances not undergoing such changes as are above indicated, and it is now known that these changes do not occur at all when due precautions are taken to exclude the living organisms. It is, indeed, conclusively proved that *bacteria* and other vegetable organisms of low type are the active agents, the real and sole cause of the ordinary fermentations, putrefactions and decompositions of dead organic matter.

The question now came, "Are the *bacteria* uniformly found in the tissues of blighting trees, themselves the *cause* of the blight, or do they simply follow the true life destroyer and are they chargeable only with the decomposition of the dead material?" To answer this question, a series of experiments was begun on the first day of July, 1880, and was continued without interruption until the fourteenth day of August following. Blight had been, and was during the time, more or less prevalent in the vicinity. On the isolated acre of ground where most of the experiments were made, there were growing ninety-four pear trees of different ages, and kinds, some in cultivated soil, others in grass; also twenty apple trees and one quince. Of the pears, seventy-two were three years old, on pear stocks. These had been transplanted in the spring of 1879, and were in good thrifty condition, making an average growth of the main limbs of about three feet. They were Bartlett's and Clapp's Favorite. The other pear trees were of bearing age, and mostly laden with fruit. Blight occurred, as commonly introduced, on four of the large trees, two Flemish Beauty, one Bartlett and one seedling, and on one only of the seventy-two younger trees.

The weather during the time, except as stated, was clear and fine, of moderate summer temperature. Much rain had previously fallen and a heavy thunder shower occurred on the last day of June. During the few first days of July the air was moist and the temperature during the day time was 90° Fahrenheit, or above. Other storms with thunder occurred July 9th and August 11th, with however, but little rainfall.

The experiments of July 1 and 10 were made by cutting pieces of diseased bark freshly taken from a blighting tree, and inserting them in the healthy three-year old trees after the manner of budding. These pieces of bark were about three-sixteenths of an inch by two-thirds of an inch. The experiments after the date just mentioned were inoculations with a sharp-pointed knife or needle dipped in virus collected as it exuded from diseased trees, and usually thinned with distilled water. As has been stated above, this exudation is composed almost purely of *bacteria*. By careful collection and frequent microscopical examinations, it was possible to be quite sure that the inoculating material contained nothing but water and the living *bacteria*. Every precaution was taken on this point as well as the thorough cleansing of the instruments used. Usually shoots of the current season's growth were selected for inoculation, and the puncture was made at various distances from the tip.

Experiments numbered thirty-three to forty-two, inclusive, were by inoculating with a needle in thirty apple trees in nursery the third season, operating in the growth of the year before, and using virus from both pear and apple. None of these were successful. Twenty-one experiments were made by simply applying virus, by means of a brush, to the uninjured epidermis of growing shoots and leaves. No one of these communicated the disease. But of the pear trees inoculated by budding and puncturing, as described *sixty-three per cent. became diseased*, exhibiting all the characteristics externally and internally, of the blight. Eighteen of these inoculated trees stood in a row of fifty-five, two feet apart in the row, and were selected here and there among those of the same kind and apparent thriftiness left without inoculating. Of the thirty-seven thus left, all save one remained free from the malady. Seventeen others of the same age and kinds growing near by continued healthy. Thus less than two per cent. of those not inoculated became infected with the disease, against sixty-three per cent. of the inoculated ones! When it is remembered that these trees were as nearly identical in kind and condition as it was, or ever is, possible to select, growing very near each other in the same soil, subject to the same treatment except in regard to the inoculations, the case seems to be settled, and an answer to our question obtained. The slight wounds made by the process of inoculation cannot be charged with the results, for similar wounds were made with a clean needle, and these rapidly healed without further injury. *The introduction of the virus introduced the cause of the disease, and the potency of the virus was quite positively due to the living bacteria.*

Of the pear trees inoculated with virus from diseased pear, fifty-four per cent. showed clear evidence of blight; while of those inoculated with virus from blighting apple limbs, seventy-two per cent. became as thoroughly infected, and as speedily died. The greater per cent. of infection from the apple virus does not necessary prove that it is more destructive to the pear. Another set experiments may and probably will indicate that it is not; but I think it may be fairly taken to show that the disease is identical in the two trees, as other indications suggest.

The four inoculations of quince with pear virus were all successful. These were upon the branches on one bushy tree whose low head was composed of thirty to forty branches similar to those experimented upon. None of these showed, at any time during the season, any trace of blight. The four diseased ones having been timely removed, the tree was freed from the disease.

Turning to the apples experimented upon, we find the results less positive. As before stated, ten of the trees were in a nursery a half mile distant from the other experimental trees. Among the hundreds of trees in the block, blight was noticed only in a single case. The ten inoculations were made in the bark of the preceding year's growth, by the use of a needle. Pear virus was used in five, but from apple in the other five. None of these were successful. The slight wounds soon healed without the least evidence of the corroding effects of the disease. No explanation can be offered for the failure of the inoculations, though it is quite possible that the results would have been different if the punctures had been made in the bark of the shoots of the current year, upon which the disease is naturally more common. Other experiments seem to favor this supposition, but in two instances the disease was successfully introduced to the bark of the trunks of much older trees. Aside from the ten mentioned, the other experiments upon apple trees were confined to the variety known as Grimes' Golden. This variety is not peculiarly subject to blight, but several young branches were trimmed away from time to time, having plain indications of the disease arising naturally. Sixty per cent. of inoculations, mostly in the bark of very young shoots, clearly communicated the disease; but except in two cases, the subsequent development of the malady was feeble compared with that upon the pears. Basing the reckoning upon the whole number of inoculations upon apple trees, thirty per cent. were successful. This, placed by the side of the sixty-three per cent. successful in pear, shows the apparent relative susceptibility of the trees to receive the disease by the process tried. But little reliance can be placed upon one set of experiments for exactness in numbers. We can only say with positiveness that blight may be induced in apple trees as well as pears by inoculations with *bacteria* from either diseased apple or pear trees.

The application of virus, known to be potent by inoculation, was made with a brush upon the uninjured surfaces of four pear trees and six apple trees. Some of the portions thus treated were bound with cloth moistened with water, to prevent the rapid dessication (sic) of the virus, but all proved unsuccessful. In a similar manner virus was applied after six o'clock P. M., upon a large number of leaves of five pear trees and two apple trees. The most perfect and apparently healthy leaves were selected, and account was kept as to the application to the upper or under surface. No certain infection followed; yet previous investigations showed me that leaves may and do suffer from the same blight, even when borne upon healthy shoots. Had some of these leaves showed infection, this would not have satisfactorily proved that the *bacteria* were able to penetrate the epidermis, for the latter might have been minutely ruptured by insects, or otherwise, and escaped notice. But the experiments do tend to show that the virus is harmless on the external surfaces of sound plants; neither does it appear to gain access to the inner tissues through the stomata, or breathing pores.

In experiments numbered nine and ten, limbs with the disease, in an active state, were cut and bound in among the branches healthy trees. No injury ensued. This was upon the first day of July, when the earth was wet and air moist from the heavy rain of the day before. Eight days afterward another storm occurred, the dead branches being left in the trees. Here again, disease might have occurred through inoculations by insects, etc.; but the failure further supports the non-communicability of the disease from without. The circumstances would seem sufficiently favorable for positive results, were these possible.

Of the total number of experiments, of which records were kept, detailed in the accompanying table, thirty-four and seventy-eight hundredths per cent. were effective, while of the whole number *inoculations* with knife and needle, fifty-two and seventeen hundredths per cent. unmistakably communicated the disease.

Thrown into tabular form the per cents., as now given, are as follows:

Number of experiments	Kinds of trees	Virus from Pear	Virus from apple	Successful inoculations	Successful experiments of all kinds
36	Pear.	54.00	72.00	63.00	38.80
29	Apple.	30.00	None.	30.00	20.69
4	Quince.	100.00	None.	100.00	100.00
69	All Kinds.	....	....	52.17	34.78

The close observations required for the work as now detailed, together with much attention given to the disease during some years, have convinced me that two popular opinions concerning the progress of this fatal disease are not founded upon facts. One is the usual supposition that destruction by this blight is very sudden; a healthy tree may, in the course of twenty-four hours, become so affected that the leaves will speedily blacken and the branch as speedily die. It is thought that a large part, or all the tree is affected at once, and death follows as it might from a stroke of lightning. I find the march of the destroyer very irregular, but always slow! The difficulty usually is to tell after the lapse of a single day whether any advance has been made, not how much. In experiment numbered seven, the progress, counting both ways from the point of inoculation, was about an inch per day for seventeen days, but this was altogether exceptional. Infection after inoculation rarely shows at all under nine days, and after this time the destroyer seldom gains more than one-fourth of an inch per day in its gradual progress through the cells. But if guided solely by the change of color in the leaves, other conclusions would be reached. Ordinarily the leaves do not become blackened until about two weeks after the infection of the bark at their insertion, except in the case of very young shoots. In number seven, just mentioned, every leaf was fresh and green in appearance at the end of the seventh day. Sometimes the bark of older portions harbors the active blight-producers all the season, while the leaves remain green until autumn. After the outer bark is killed throughout some feet in length of trunk or large limbs, new layers of bark may be formed within, and the tree recover, possibly without the cultivator being aware at all of the crisis through which it has passed. In old limbs and trunks covered with rough bark, the daily progress of the disease cannot be observed; but this is not usually the case in those parts still covered by the smooth epidermis. The color changes, a watery appearance is presented, and especially in dewy mornings, little beads of white or tinted gummy exudations may be seen. If the surface is really wet, this material is dissolved and spreads over the surface, or runs down the bark, often dripping from projecting points. When these characteristics fail to mark the progress of the destroyers, thin shavings from the outer bark quickly and unmistakably give the information. After the disease is well started, the watery appearance of the freshly cut tissues is recognizable within a few hours after invasion by the *bacteria*; and after as many days, decided change of color takes place shown the death of the cells. The brownish change extends in projecting points and streaks into the healthy tissues; these soon spread laterally until they meet, either entirely around the part, or as not unfrequently happens, only on one side, in the latter case forming a roundish patch or an elongated strip of dying bark. The more or less circular form seems to be especially common on older parts, as the trunk of trees; but the outlines are always irregular, sometimes decidedly so. After a time the further spread of the disease is often checked, or altogether stopped, by natural causes not yet known.

The second popular notion believed to be erroneous, is the influence of thunder storms in accelerating the progress of the disease. So prevalent is this impression, I am disposed to think the appearance really does give support to the idea, though my own observation is not at all conclusive upon this point. But by carefully noting the exact progress of the malady in the bark as already indicated, I have not been able to detect any variation corresponding to the changes of meteorological conditions. Doubtless there are some such variations; but assuredly not such as are appreciable on account of passing

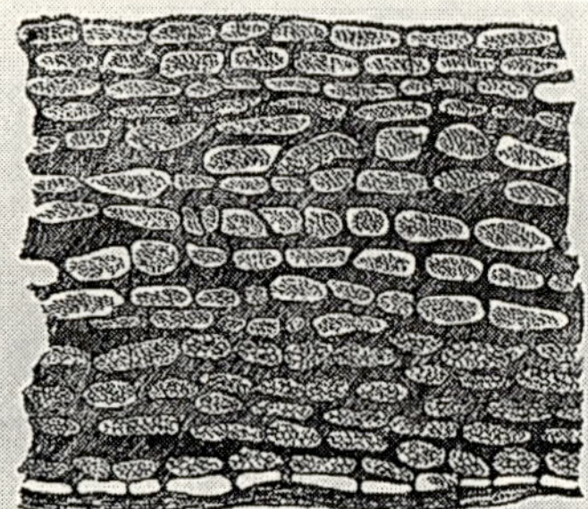

Fig. 1. Cross section of healthy bark taken August 1. The cells are full of starch. Magnified 125 times, linear.

storms. Even winter weather does not offer complete immunity to the trees. During two winters, I have watched the slow progress of the disease in the branches of the same tree, and am positive there was no mistake as to the nature of the phenomena. In the first case[11], the gradual destruction was continued into and through the spring; in the second (1878), further progress immediately ceased with the expansion of the buds. If transient thunder storms really have any effect, it is upon the leaves only; and it probably is true that when these, or the shoots upon which they grow, are already diseased, the change in color is greater when the air is saturated with moisture than at other times and under other conditions.

The organism to which we attribute the death of our fruit trees is so minute that a magnifying power of two hundred times across is necessary to see it at all, and one thousand diameters is required for careful study. Only the objectives of sharpest and finest definition give satisfactory results. My own mistake, recorded in the Transactions of the Illinois Horticultural Society, 1877, p. 114, by which these minute bodies were equivocally referred to the *spermatia* of some fungus, shows the want at the time of a proper glass. The wide-angled objectives of recent manufacture are expensive instruments, but are immensely superior for this kind of work.

Fig. 2. Cross section of diseased bark from same tree, at same time as Fig. 1. Magnified 125 times.

During the course of the development of the minute creature of which we write, it assumes various shapes, and those of different forms may ordinarily be seen together in the field of the microscope. The most characteristic form, and that most abundantly seen, consists of two oblong joints with obtusely rounded ends, attached close together by a broad base, end to end. Their transverse diameter is about one-thousandth of a millimeter (.0000394 inch), and the length of the two joints is three-thousandths of a millimeter (.000118 inch). They are comparatively thicker and shorter than *Bacterium*

[11]Transactions Illinois State Horticultural Society, 1877, p. 114.

tenno, so common in putrifying (sic) substances. Their motions are less rapid. They slide forward with a slightly undulating motion, they turn over and on end, but never glide with rapidity across the field.

It is exceedingly difficult to determine specifically these minute creatures by size and form, or what are termed morphological characteristics alone. This one appears to me distinct from any that have been described, but I do not undertake at present to decide this question. It is quite possible the active atom, which we find in the diseased tissues, is identical with a common, omnivorous little agent,[12] which converts sugar, amylaceous matter, lactic, tartaric, citric malic and muric acids and albuminoid substances into carbon dioxide, butyric acid and hydrogen, whenever and wherever the conditions permit. Should this prove to be the case, it would not necessarily, if presumably, invalidate its active agency in producing the disease of the pear and apple tree; but it might render less hopeful the discovery of preventive treatment. So far as my observation goes, the allied organism found by myself in specimens sent me from Michigan of peach tree afflicted with the "yellows," is, in form and size, very similar to that of the pear blight; but, according to my own measurements and those of Dr. H. J. Detmers, who possesses a magnificent microscope furnished with the very best of Tolle's objectives, the *bacteria* of the peach yellows is, in its characteristic form, one-thousandth of a millimeter in transverse diameter (the same as that from the pear) and three and a half-thousandths of a millimeter long; hence, slightly longer than the pear organism. (Pear, .0000394 inch by .000118 inch; peach, .0000394 inch by .0001377 inch). The peach organism is more commonly more than two-jointed, under the sharpest definition appearing four-jointed nearly as frequently as two. Neither of them is *Bacterium tenno.*

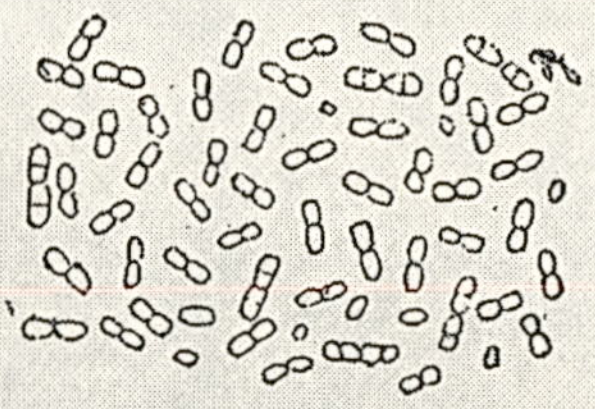

Fig. 3. Bacteria, causing blight in pear and apple trees. Magnified 1,000 times, linear.

[12] *Vibrio butyrique*, Pasteur. *Bacillus amylobacter*, Van Teighem. These are now believed to be identical.

The most conspicuous change that can be observed, by the aid of the microscope, in the tissues affected with the blight, is the disappearance of the stored starch. *The cell walls are not dissolved or altered, in any way,* except by the staining, which sometimes takes place through oxidation, in the later stages of the disease. This is an important point, and though at variance with many published statements, has been established by numbers of microscopical observations devoted to this special problem. If there be any exception, it is confined to the thin walls of very young cells, which soon shrink, and become by drying much distorted. But unless the *bacteria* do dissolve the cellulose of the cell wall sufficiently to make a passageway for themselves, it is impossible for them to gain entrance to such cells as they infest in their adult form. So it must be likewise impossible, in the same condition, for them to spread through tissues as we find them doing. They are assuredly not carried in the circulation of the fluids of the tree; where this is most active they are not found at all, and their slow dispersion from the point of beginning, in all directions through the starch-bearing cells, is too uniform for such interpretations. It is also readily enough proved that the tissues primarily affected by blight have no open pores in their cell walls, such as occur in those of older wood, etc.; neither are there ducts or other channels for their passage. The circulating fluids are filtered most thoroughly in their passage through the walls of each cell; it is impossible for solid particles large enough to be seen by our best microscopes, to pass through the plant tissues. The speculations often made, of ordinary spores of fungi being absorbed by the roots and carried by the watery currents in the plant, are wholly erroneous. Any tyro in microscopy can disprove the too-common assertions of this kind.

Accepting[13] Nägeh's theory of the molecular construction of the cell wall, we can only understand how the organisms pass from cell to cell in their deadly work, by supposing their germs are less than the molecular openings, hence ultra-microscopical, and that in this condition they pass the cellulose barrier and develop into the forms we observe within: or that in some condition they are so plastic, that, amœba-like, they are able to creep through the narrow spaces between the molecules, just as water penetrates the wall. It is

[13] Sach's Text-Book of Botany, Eng. Trans., p. 588.

well known that the spores of many parasitic fungi, e. g. *Peronospora infestans*[14] causing the potato -rot, send their germ-tube, narrowed to exceeding fineness, though still visible, through the epidermis of the leaf. Through this delicate tube the contents of the spore is transferred to a swelling formed at its apex. Thus the essential portions of the spore gain admission to the inner tissues, through which is able to thrust its ramifying threads in every direction. The minute opening in the outer wall of the epidermis becomes closed and all indications of the entrance become obliterated.

If anything like this occurs in the penetrations of our *bacteria*, it has not been observed, and is probably on too small a scale be made out.

After a cell has been invaded by the moving particles, the first thing noticeable is the agitation of the starch granules. They visibly tremble, they swing minutely to and fro, bumping each other very gently in the confines of the cell. Then dissolution begins, the grains lose their characteristic form, and finally entirely disappear. The protoplasm of the cells is not directly affected. By staining with carmine it may be found in abundance after the stored starch has all disappeared, and after the branch shows by external signs the death of the parts. Probably carbonaceous compounds held in solution, as sugar, organic acids, etc., undergo the same fermentation and destruction as the starch; but the microscope gives no evidence of this. So far as we can *see*, the starch is the only thing upon which the *bacteria* live. In so living, they cause the death of the other portions partially by starvation, partially, no doubt, by the action of injurious substances into which the amylaceous matter is converted.

There is absolutely no trace of other fungus growth in the tissues until after death has taken place, neither have any been found for several days after this in twigs or diseased parts of trees left in the meantime out of doors. In culture baths, ordinary moulds soon appear on the sections of infected parts. If the death of the trees were due to a species of the larger parasitic fungi, like that of the potato rot or of the wheat rust, surely it would have been found by some of the trained eyes that have patiently sought for it. Obscure as they are, none are too much so for the modern microscope, and its mysteries not too inscrutable to be past finding out.

---

[14] DeBary, Morphologie und Physiologie.

In very young tissues, such as the tips of apple tree shoots, all part except the epidermis seem to be equally invaded, but in older limbs the bark parenchyma, or the outer layer of living bark, is the first and usually chief seat of the disease. The bast, composing the inner fibrous layer, is not affected. Not unfrequently a close layer of bast cells divides healthy from diseased and dying parts. Sometimes invasion through a layer of bast is gained by way of medullary rays which pass from the wood outward through the fibrous bark, aud (sic) whose cells are often stored with starch. Contrary to the usual opinion, the cambium, or new layer of thin walled, growing cells just outside of the wood, is by no means the seat of the disease. Often when the whole thickness of the bark perishes, the cambium retains its vitality, and sometimes forms a fresh layer of new bark beneath the old dead coating. In this case the limb or tree may survive without severe, apparent injury. In summer, such a change may take place while the leaves remain green upon the infected branch, showing nothing of their narrow escape from destruction. The wood of a young limb is often discolored above the diseased part by the ascending water colored in its passage through dying and brown portions. The staining is especially noticeable in the fibro-vascular bundles (woody fibers) which turn out into the leaf stalks, and which are the chief water carriers to supply the loss by transpirations from the leaf surfaces. This stained wood may be still healthy in every particular, and continue its proper functions, upon the recovery of the really diseased parts. There is no evidence of the progression of the disease in the wood through which the chief movement of the water on its ascent from the roots to the leaves takes place. The leaves, however, are usually infected from the diseased bark from which they arise. The *bacteria* work their way through the tissues of the leaf stalk, and spread throughout the parenchyma (soft tissue) of the blade. It is also probable that infection starting in the leaf may spread downward to the stem when conditions are favorable, but infected leaves do sometimes die and fall from their attachments without communication of the malady to the cells of the branches. The dying leaves become smeared, especially during dewy nights, with the exuding *bacteria*. When dry they form a varnish-like coating upon the leaf surfaces, in which condition they retain their vitality for an undetermined length of time.

CHEMICAL CHANGES

My colleague, Professor H. A. Weber, kindly determined for me the kind of fermentation which takes place in the diseased bark. He found, first, that carbon dioxide is abundantly given off. Bark in which the fermentation is active put into water in a stoppered bottle, quickly shows the forming gas. In a few hours the pressure resulting from the accumulating gas forces the frothy liquid out of the bottle in a slow but steady stream. Secondly, butyric acid was determined as a considerable product of the fermentation. This is one of the fatty acids, perhaps best known from its occurrence in butter; it has been heretofore mentioned as one of the products, with carbon dioxide and hydrogen, of a peculiar fermentation of the hydro carbons (sic), such as starch, sugar, various organic acids, etc. It is now well understood that such fermentation is produced by *bacteria*, either by a special species for each substance, or perhaps one and the same for all of these organic compounds. Cellulose (cell wall substance) also undergoes, when the conditions are favorable, the same kind of fermentation. A brief account of investigations upon these peculiar transformations will directly contribute to our information upon the newly observed fermentation in the living cells attacked with blight.

Until about the beginning of our own century it was generally understood that, as soon as the influence of life was withdrawn, organic bodies began to decompose spontaneously, or without any external cause. It was supposed the vital forces had built up the compounds in opposition to the action of chemical and physical laws, and that the simple cessation of the former was sufficient to permit the latter to retransform the complex life-built bodies simpler ones, or to the original elements. Willis[15] (1659) and Stahl (1697) held that an organic compound *in a special state*, was necessary to set up fermentation, which was accomplished by transmitting to the molecules of the fermentable body a peculiar motion which new arrangements of the elementary substances were affected. In 1680 Leuwenhœck first detected yeast in fermenting beer, but knew nothing of its nature; and it was not until about 1837 that the independent investigations of Latour, Schwann, Kützing, Turpin,

[15]The volume of the "International Scientific Series," by Shützenberger, "On Fermentation," contains a condensed history of the subject.

Mitscherlich, and others, proved yeast to be a mass of living vegetable cells. The general result of the investigations of these scientists gave support to the idea, first correctly formulated by Latour, that the *yeast cells caused* such fermentation as occurs in the manufacture of beer and bread (Ann. Chim. Phys., 2nd series vol. 68). This theory, supported as it was by so many careful investigations, ought, it now seems, to have received the assent of scientific men. But this was not the case. From the very commencement of the discussions upon them, Latour's conclusions found a powerful opponent in one whose name at the time was a decide authority in chemistry, and whose influence is still strongly felt upon this and many other questions relating to the practical affairs of life. Justus Liebig had adopted, through Gay-Lussac, the explanations of Willis and Stahl, already noticed, and, by many brilliant experiments and illustrations, he extended and satisfactorily expounded the theory of fermentation through the communication of peculiar atomic motion in the elements of the fermentable compounds. The important experiments of Appert upon the preservation of animal substances, and of Gay-Lussac upon the freedom from fermentation of the juice of grapes excluded from the air, were cited and explained, and withal a very philosophical and seducing interpretation given of the obscure phenomena taking place in the transformations of what were supposed to be unstable organic bodies. Liebig, however, recognized the necessity of *some external cause to introduce* fermentation in a body in which it had not already begun. It could not arise spontaneously, and in the absence of an exciting cause decomposition of even the least stable organic bodies was known to be indefinitely postponed. He claimed that when the molecular motion was once excited, fermentation continued without further external agency, the motion already existing being communicated to the particles adjoining, from the latter to the next, and so on until the transformation was completed. The oxygen of the air was believed to be the most common exciting agent, hence it was simply necessary to exclude the latter, as in the ordinary processes of canning, to preserve fermentable substances. Liebig's theories gained very general scientific and popular acceptation, and have been very slowly, apparently reluctantly, abandoned. He defended them with much vigor against the conclusive experiments of Pasteur, which beautifully confirmed and extended the supposition of Latour.

So late as 1870, the great German chemist published a lengthy memoir (Ann. der Chemic und Pharmacie, vol. 153, p. 1,) upholding his mechanical theory of fermentation against the physiological one renewed by Pasteur; but in this he makes concessions which are almost equivalent to an avowal of defeat. Pasteur replied in 1872, and Dumas showed by ingenious experiments, conducted with his consummate skill, that it was impossible to verify the consequences of the Liebig theory. Tyndale, in —, published an account of his wonderfully precise and beautifully conclusive experiments, showing that atmospheric oxygen freed from living organisms has no power to excite fermentation, putrefaction or decay.

There are still differences of opinion among scientists concerning these phenomena; but we may receive it as definitely settled that fermentations producing alcohol are excited and continued by the physiological action of yeast, and that those producing acetic, lactic and butyric acids, and the analogous transformation giving rise to ill-scented gases and known as putrefactions, are similarly induced and sustained by *bacteria.* Both of these groups of organisms have been variously classified as fungi, as algæ, or even as animals; but this does not indicate disagreement upon their physiological activeness and effects. The latter constitute the so-called "disease germs," to which are attributed by many the contagious and infectious diseases of animals, and to which we now refer diseases of plants.

I now append a short sketch of investigations and interpretations that have been made, more closely bearing upon the special subject before us.

In 1850, Mitscherlich[16] communicated to the Academy at Berlin his studies upon the fermentation of cellulose, and expressed the opinion that it was due to a particular organism which he found in the fermenting material. Pasteur[17], in 1861, studied the butyric fermentation of sugar and calcium lactate, and concluded that the transformations were produced by a special infusorium "composed of little cylindrical rods, rounded at the extremities, usually straight, either isolated or united in a chain of two, three or four joints,

[16]Comptes Rendus, t. 88, p. 205: t. 61, pp. 156 and 436, t. 65, p. 513.

[17]Comptes Rendus, t. 52, p. 344.

and even more." This he placed in the genus *Vibrio*. In 1865, Trecul[18] found in macerated vegetable substances, a living organism, which he believed arose spontaneously from a direct transformation of the protoplasm in the closed cell. Having assured himself that it stored starch within itself in certain stages of its development, he named it *Amylobacter*. Twelve years afterward Van Tieghem[19], after careful study, transferred the organism to the genus *Bacillus*, retaining, however, Trecul's name for the species. The scientific name now became *Bacillus amylobacter*. The *savant* last mentioned as studying the little creature, satisfied himself that it could and did penetrate the cell walls of plants in the adult state, thus overthrowing as he thought, the idea of spontaneous origin. Here, however, the battle still wages. Trecul[20] holds his first opinion, criticising, and with apparent justness, Van Tieghem's proofs. In June, 1879, Prazmowski[21], of Leipzig, announced the discovery that this *Bacillus amylobacter* produced, as a product of fermentation, butyric acid, and concluded from this that it was the same organism as that described by Pasteur and classed in the genus *Vibrio*. Van Tieghem had, before Prazmowski's announcement, suspected this identity, and now by further investigation confirms the opinion.

It therefore appears probable that the butyric fermentation of at least all dead carbonaceous compounds, soluble or insoluble, is directly or indirectly produced by one specific organism, a very minute, living, moving, rapidly multiplying, wonder working creature. Does it now extend its dominion, and, conquering the life forces of our trees, seize upon their treasured products and ruthlessly appropriate them to uses of its own? Present information leads me to think the organism of the blight is specifically distinct, as well as that of the yellows; but this cannot now be definitely decided, neither is it important that we should hasten the verdict, the life history of the creature, its mode of

---

[18] Comptes Rendus, t. 52, p. 344.

[19] Bulletin de la Societé botanique, March 23, 1877. Comptes Rendus, t. 88, p. 205.

[20] Comptes Rendus, t. 88, p. 401 (1879).

[21] Comptes Rendus, t. 89, p. 5 (1879).

action, our methods of preventing its ravages, are much more important than the name by which it is known.

Whether the species is described or not, this proof of disease plants produced by bacteria-like organisms must contribute something towards the solution of the germ problem in diseases of animals. By careful experiments to determine the exact facts of its influence and operation in the plant cell, where the conditions can be controlled and manipulations can be made much more favorably than in animals, we may possibly gain for human pathology very important scientific and practical results. Thus science aids science, and important help comes from unexpected quarters.

REMEDIAL MEASURES

No claim is here made to the discovery of such prophylactic or remedial methods as shall change fruit growing from failure to success, or, indeed, shall materially modify the methods of the best cultivators in any way. While something is to be advanced in the way of improvement in the means usually adopted to save our pear trees, the information already given may be of more service in saving us from crude hypotheses and useless labor, fighting in the dark an unknown foe.

There is a common but often mischievous supposition that the attacks of vegetable as well as animal parasites are always preceded by an enfeebled condition of the host or affected plant. This is true enough in some cases, but by no means uniform enough to called a rule. Indeed, so far as the operations of parasitic fungi are concerned the rule is that the special condition of the supporting plant as to feebleness or vigor has no effect, and exceptions occur not very unequally upon both sides. External conditions, as of temperature, moisture, light, ventilation, etc., are always controlling influences in plant growth, whether of the higher or lower organized classes; but the effect of these, singly or combined, is by no means the same, even upon species, belonging to the same group, much less so upon widely different classes. The same variation in temperature may be injurious to the vital processes of a leafy plant, and beneficial to those of a fungus parasite in the tissues of the former, and the reverse. So it happens that certain fungus parasites may, like certain insects, be much more destructive in one climate than in another. There is something, too, in what we call constitutional hardiness or capability to withstand vicissitudes, and this applies to varieties as well as orders and classes. But the same peculiarity which prevents the operations of one parasite may predispose the plant to the attack of another, – even, perhaps, of an allied species.

Aside from the selection of varieties which have been proved by trial to be least affected by blight, the preventive measures here recommended are such as would, without the blight, be best for the growth and healthfulness of the tree. No one need hope by starvation as in grass sod – or severe mutilation, as by root pruning – to obviate the tendency to blight, unless, in peculiar circumstances, these are really favorable in the direction indicated. In ordinary soils good culture, managed so as to secure thriftiness of growth in the early part of the season and thorough ripening of the wood in autumn, must be best everywhere. Everything tending to lessen the depredations of ordinary insects of every kind, whether leaf eaters, fruit gnawers, trunk borers or anything else, will tend to prevent the introduction of blight. The contagion is not usually, if ever, carried by the air, for it is conclusively established that certain species of *bacteria* are not commonly disseminated in this way, and the viscid character of the species causing blight teaches us it is not likely to be found in the air. But however abundant in the air currents or upon the uninjured surfaces of the trees, it is not probable that it gains entrance to the tissues without help. Anything making punctures or wounds, even though very minute ones, furnishes this help and renders blight possible. The pruner's knife need not assist in the same way, but if it does not, scrupulous care must be exercised to avoid it. A knife used upon diseased limbs is well furnished with bacteria and in good order for inoculations. Wounds left exposed are also dangerous. Washes of different kinds have been apparently useful and we may now understand why. Without sufficient experience for positive tests, I recommend washing the trunks and larger limbs once or twice per year with strong lye from wood ashes, crude potash or a solution of the concentrated lye of the shops, for the purpose of making these parts clean and smooth. Exposure will in this way be reduced by destroying insects and their harborers, etc., by rendering less likely any cracks in the bark by the expansion of the stem. Linseed oil has been highly recommended and as severely condemned; but if applied after the above to trees, free at the time from infection, the effect must be in the right direction. By forming an impervious but elastic coating, it resembles the natural epidermis, the best protection the tree possesses. Petroleum is bad, for it kills the outer stratum of bark, causing cracks and roughness. Lime and sulphur solutions or mixtures do no injury and may be beneficial. A dilute solution of carbolic acid is still more likely to prove a protecting wash.

Whether it is possible to apply to the soil, to be absorbed by the roots, or to introduce to the circulation in any other way, preventive or remedial substances is a question to be solved by further experiments. Probably not. A tree can not be doctored like an animal by forcing it to swallow medicine. Materials presented to the root may or may not be absorbed. If the former, they may or may not spread throughout the tissues, or may not spread in such manner as to reach the diseased parts. The roots do not blight except in cases of contagion from the trunk, hence, so far as they are concerned, we do not need protecting substances about them. So far as my examinations show, this is also true of the peach yellows. The roots sent me were certainly free from disease, while the bark and pith of the part above ground were full of *bacteria*. It has been supposed that the chief seat of this disease is in the roots. I cannot say these are not sometimes infected, but am well convinced that they are not always so when other parts are severely suffering.

I ought here to express my decided opinion that neither the blight nor the yellows is due to any lack of special nutrient substances in our soils. After careful attention to this popular idea and abundant opportunity for observation in regard to the first mentioned disease, I find nothing in its support. The physical differences in soil, especially in relation to drainage, and the ability to withstand drouth, have important bearings on the general healthfulness of the trees and indirectly may affect the tendency to blight; but that there in the Western States, either upon new or long cultivated soils, any special poverty of mineral ingredients necessary to the proper food of the trees, corresponding in any way to the ravages of blight, is certainly not proved, and for anything I can find out is wholly untrue.

If we now ask what may be done after infection has taken place in the pear and apple, there is but one remedy which can be confidently depended upon to check the progress of the disease. This is the old one of removing the infected parts. There has been much difference of opinion as to the usefulness of this procedure, but in the light of these investigations, this difference has certainly arisen from want of knowledge as to the evidence of infection in its earlier conditions, and from failure, through inattention, to really accomplish the work undertaken. He who cuts away only the parts already cleared, or who uses a knife or saw smeared with material from diseased portions, must not wonder that his labor is in vain. Neither should we be surprised if the disease spreads from an unprotected wound made in healthy portions with a carefully cleaned instrument. It is very plain that, to be successful, the really infected parts, whether conspicuously changed in color or not, must be excised, and that by some method not likely by its very process to re-communicate the infection.

It is true this cure must sometimes be severe, but the proved slowness of the progress of the malady, together with its positively local character in the tree, are good reasons for thankfulness that we have a certain remedy, and ought to give us confidence in its use. If the alkaline washes heretofore commended are applied once or twice during the season, the resulting smoothness and freshness of the bark will aid materially in early detecting the presence of the *bacteria*. In young limbs this can always be done as before indicated, not by waiting for the leaves to reveal the destruction already accomplished, but by evident changes occurring in the bark. In the absence of wounds or bruises upon older parts, as well knotty places from bad pruning and other causes, the thickness the outer layer of corky substances provides an excellent protection; hence, in well grown, well kept trees, the parts most liable to receive disease-producing inoculation are such as quickly show the effects of the destroyer. Among such trees, careful examination once in two weeks, especially during the months of June, July and August, and as careful, well-managed use of the knife, may not insure freedom from serious loss, but certainly will, in most cases, materially and hopefully reduce the ravages of this heretofore mysterious disease.

It is reasonably probable that some antiseptic application up the surface, at least after removal of the epidermis, or layer of dead bark, may be found an effective remedy; but I am not experimentally satisfied upon this point.

Carbolic acid (pure) five parts, to one thousand of water, kills the *bacteria*; other similar substances are known to act equally well much more diluted; but this can readily be obtained, and may prove the best practical remedy. Professor Turner, of Jacksonville, Illinois, thinks he has during many years prevented the spread of the disease from infected parts by painting the surface with turpentine, in which, for convenience, lamp-black had been stirred.

After a limb has been cut away, the fresh wound should be washed with the one-half per cent. carbolic acid solution, and immediately covered with linseed oil, oil and lead paint, or shellac dissolved in alcohol. If a spot upon the trunk or a large limb shows infection, the bark should be sliced away deep enough to remove the invaded portions as indicated by the appearance of the tissues, being careful not to disturb, unless necessary, the liber and cambium, the surface washed with carbolic acid solution and painted as before, or bandaged with cloth or stout paper, made impervious by wax, oil or parafine (sic). An excellent bandage may be made by applying, with a brush, a coat of linseed oil upon sheets of manilla (sic) wrapping paper, and cutting these, when dry, into strips two inches wide. The strips may be used just as a surgeon uses bandages of muslin, by winding spirally over the wound. Mr. Ricketts, of Newburgh, New York, reports, after many trials, good success by paring off the diseased bark and binding tempered clay upon the exposed tissues.

Upon apple trees, the blight is more commonly confined to the young shoots, and upon these the leaves quickly indicate the presence of the disease. Prompt removal of the shoots, taking care cut below the infection, even without further treatment, must be advantageous. When, however, larger limbs and trunks are attacked, the method proposed above is required.

It can do no harm, and may be useful to burn the severed, diseased parts; but infection from such portions thrown upon the ground cannot be very direct, if likely to occur at all, under ordinary circumstances. The knife may be disinfected, from time to time, by dipping in a strong solution of carbolic acid, carried for the purpose, or by passing it through a flame in such manner that the surface may be momentarily highly heated without injuring the temper of the blade. An alcoholic lamp furnishes facility for this. Scouring thoroughly in the earth would probably, but not certainly be effective.

According to the current opinion concerning the slow contagion of the peach yellows, it may be practically possible to "stamp out" the disease from any given locality. This will largely depend up the specific nature of the parasitic organism, and whether or not lives upon anything else than the tissues of the peach tree. This process does not appear feasible in regard to the blight of the pear and apple; yet, further investigations may reverse present probabilities. Stranger things have rewarded research than the discovery of some well known plant which, by serving as nurse to the minute parasite in an essential stage of its development, is indirectly responsible for the ruinous disease of the pear. Should, for example, conviction of this kind rest upon the Lombardy poplar, how widely different the cure from our present proposals!

We shall conquer when we know how. When we, with open eyes and unstopped ears, as true students of Nature, acquire the knowledge within the possibilities of our reach, we shall be able to rightfully assert our royal authority, and to have effectual domination over every living thing that moveth upon the earth. The following table gives in detail the number of experiments tried of which records were kept, the date when performed, the kinds of trees operated upon, the source of the infecting material (virus), the method of applying the latter, the date when the first evidence was noticed of the successful result of the inoculation, the number of days elapsing between the time of the inoculation and the first noticed evidence of infection, the number of days elapsing between the date of first visible evidence of the infection and its spread, counting both ways from point of inoculation of four inches, and some miscellaneous matter. In the fourth column, the initial letters P. and A. denote pear and apple, the inoculating material having been taken from these trees in various stages of blight. In the fifth column, the word "budding" denotes that a small piece of diseased bark was inserted in the healthy tree; the word "needle" is meant to indicate that the healthy bark was simply punctured, either with needle or the sharp point of a small knife, previously charged with virus; the word "brush" denotes that the virus was applied to uninjured parts of healthy trees. The sixth and seventh columns show, in different ways, the apparent time required for the small amount of the inoculating material to effect change in the tissues sufficient to be observed by external characteristics, and the eighth column gives some idea of the rate at which the disease progresses, mostly judged in the same manner. Nearly all, except numbers 33 to 42 inclusive, were in shoots of the current year's growth. In old trunks, etc., the spread of the disease is much slower, but irregular.

EXPERIMENTS IN DETAIL.

No. of Experiments	Date, July and August, 1880	Kind of Tree.	Virus	Method.	First Evidence of Infection.	Days after Inoculation	Days Progressing 4 Inches	
1	1	Bartlett Pear	P	Budding	17	16	8	.....
2	1	"	A	"	21	20	5	.....
3	1	"	P	"	...	...	...	Bound with cloth.
4	1	"	A	"	...	...	...	.....
5	1	"	P	"	...	...	...	.....
6	1	"	A	"	22	21	9	.....
7	1	"	P	"	9	8	4	Diseased 18" in 17 days.
8	1	"	A	"	15	14	9	.....
9	1	Clapp's F. Pear	P	Limb	...	...	...	.....
10	1	"	A	"	...	...	...	.....
11	10	Bartlett Pear	P	Budding	19	9	3	.....
12	10	"	A	"	22	12	5	.....
13	10	Anger's Quince	P	"	23	13	...	.....
14	10	"	P	"	22	12	...	.....
15	12	Bartlett Pear	P	Needle	...	...	...	Bound with cloth.
16	12	"	P	"	22	10	3	.....
17	12	"	P	"	...	...	...	.....
18	12	Clapp's F. Pear	P	"	...	...	...	...
19	12	"	P	"	24	12	...	Bound with cloth.
20	12	"	P	"	20	8	7	.....
21	12	Bartlett Pear	A	"	23	11	4	.....
22	12	"	A	"	24	12	5	.....
23	12	"	A	"	...	...	...	.....
24	12	Clapp's F. Pear	A	"	21	9	3	.....
25	12	"	A	"	22	10	1	Bound in cloth; young
26	12	"	A	"	...	...	...	shoot.
27	12	Bartlett Pear	P	Brush	...	...	...	.....
28	12	"	P	"	...	...	...	.....
29	12	"	P	"	...	...	...	.....
30	12	Clapp's F. Pear	P	"	...	...	...	.....
31	12	"	P	"	...	...	...	.....
32	12	"	P	"	...	...	...	.....
33	13	Apple	P	Needle	...	...	...	In bark of last year.
34	13	"	P	"	...	...	...	In bark of last year.
35	13	"	P	"	...	...	...	In bark of last year.
36	13	"	P	"	...	...	...	In bark of last year.
37	13	"	P	"	...	...	...	In bark of last year.
38	13	"	A	"	...	...	...	In bark of last year.
39	13	"	A	"	...	...	...	In bark of last year.
40	13	"	A	"	...	...	...	In bark of last year.
41	13	"	A	"	...	...	...	In bark of last year.
42	13	"	A	"	...	...	...	In bark of last year.
43	17	Grimes' Apple	P	"	...	...	...	.....
44	17	"	P	"	...	...	...	.....
45	17	"	P	"	...	14	18	.....
46	17	"	P	"	...	11	16	.....

No. of Experiments	Date, July and August, 1880	Kind of Tree.	Virus	Method.	First Evidence of Infection.	Days after Inoculation	Days Progressing 4 Inches	
47	17	"	P	Brush	...	...	...	.....
48	17	"	P	"	...	...	...	.....
49	17	"	P	"	...	...	...	.....
50	17	"	P	"	...	...	...	.....
51	27	"	P	Needle	...	16	...	Apex died.
52	27	"	P	"	...	12	...	Two inches died.
53	27	"	P	"	...	17	...	Three inches died.
54	27	"	P	Brush	...	...	...	Leaves, upper surface.
55	27	"	P	"	...	...	...	Leaves, lower surface.
56	27	"	P	"	...	...	...	Leaves, upper surface.
57	...	.....	P	"	...	...	...	Leaves, lower surface.
58	...	.....	P	"	...	...	...	Leaves, upper surface.
59	31	Bartlett Pear	P	"	...	...	...	Leaves, upper surface.
60	31	"	P	"	...	...	...	Leaves, lower surface.
61	31	Clapp's F. Pear	P	"	...	...	...	Leaves, upper surface.
62	31	"	P	"	...	...	...	Leaves, lower surface.
63	31	Grimes' Apple	P	"	...	...	...	Leaves, upper surface.
64	31	"	P	"	...	...	...	Leaves, lower surface.
65	31	Anger's Quince	P	Needle	11	11	15	.....
66	31	"	P	"	14	14	16	.....
67	5	Grimes' Apple	P	"	...	...	...	.....
68	5	"	P	"	14	19	...	.....
69	5	"	P	"	...	...	...	.....

Joseph Charles Arthur

Joseph Charles Arthur was born in Lowville, New York, on January 11, 1850.[1] He developed an interest in plants during his childhood when his family moved to the prairies of Iowa, but he received little encouragement from his parents to further this interest. They favored a more lucrative profession for their son, arguing that "botany could not . . . bring in enough to keep a cat alive." He could not be dissuaded, however, and entered Iowa Agricultural College in 1869 where he was part of the college's first graduating class in 1872.[2] This was the beginning of a long line of firsts for Arthur. In 1878, he presented the first Masters thesis (not on plant diseases, but *On the Structure of Echinocystis lobata*, a member of the cucumber family) to the Iowa Agricultural College.[3] In 1886, he received the first doctorate awarded in a scientific field by Cornell University, which was also the first doctorate awarded in the United States for research on plant diseases, for his work on fire blight. He was the first botanist at the agricultural experiment station in Geneva, New York, and, in fact, the first staff botanist at any state experiment station. He also was credited by his contemporaries with doing the first institutional, systematic study of plant diseases in the United States while at Geneva.[4] Then later, he was the first botanist at Purdue University. Perhaps most significantly, Arthur was the first scientist to be known as a "phyto-pathologist."[5]

After graduating from Iowa Agricultural College with a master's degree in 1878, Arthur had difficulty finding a suitable job. While traveling to New Jersey to take a job in a greenhouse, he stopped in Baltimore to look around newly opened Johns Hopkins University, a school modeled on the German university system. He was unexpectedly ushered in to see President Daniel Gilman who knew Arthur's name and reputation via the great botanist and teacher Charles Bessey, Gilman's friend and Arthur's mentor at Iowa. Gilman offered Arthur a fellowship on the spot, enticing him with the news that Harvard's William Farlow, the most famous botanist in America, was coming to deliver a series of lectures. Arthur accepted and set about retrieving his luggage from New Jersey.[6] He did meet Farlow and, in fact, went to Harvard for a summer course after his fellowship in Baltimore ended. After Harvard, Arthur became a botany instructor at the University of Wisconsin from 1879 to 1881 and then briefly at the University of Minnesota.[7]

In 1884, Arthur met Dr. E. L. Sturtevant, director of the New York State Agricultural Experiment Station at Geneva, at the annual meeting of the AAAS and was offered the position as station botanist.[8] When Arthur arrived at Geneva, he found the station to be an agricultural institution of limited scientific resources. He did have access to Dr. Sturtevant's microscope which was "covered with all sorts of brass instruments, scapels, needles, etc. with ivory handles," but aside from Sturtevant's baroque glass, the station was ill-equipped.[9] Arthur's worktable was a board placed over the radiator by a window. He had a few test tubes, cotton, and simple apparati; however, the facilities were crude by the standards he had seen at Harvard, Johns Hopkins, or Iowa Agricultural College.[10]

What Arthur did have which gave him an advantage over many of his colleagues elsewhere was control over his research. "I find I am free to plan and execute any investigation I see fit," Arthur confided to Bessey. He was not sure exactly what he wanted to do, but he had "concluded to look into the life and habits of some of the plants that cause diseases of cultivated crops."[11]

Because upstate New York was a center of tree fruit production, it was not a peculiar decision when Arthur quickly settled on fire blight as a research topic. He accepted the recent proof of the bacterial origin of the disease, but knew more work needed to be done to establish the fact beyond doubt. Arthur saw his mission as a follow up to the disease transmission experiments by Burrill and wrote:[12]

> The experiments of Professor Burrill showed that the disease alluded to was invariably accompanied by a specific form of bacteria. . . . Although from these and subsequent investigations, the theory has been quite generally accepted

> that bacteria are the cause of the disease, no rigid proof of it has yet been brought forward.

The first year, 1884, he repeated Burrill's experiments and demonstrated quickly and irrefutably that infection occurred from an infusion, whether created by placing diseased wood in water to create a bacterial solution or by transferring exudate straight from a blighted twig into a healthy shoot.[13]

The next year, in the spring of 1885, Arthur began to expand on Burrill's work. He embarked on a series of "pure culture" experiments at the suggestion of A. N. Prentiss of Cornell University. Arthur began to separate the bacteria from "the juices which accompanied them."[14] He created what he referred to as "artificial" or "fractional" cultures through two series of six serial transfers with sterilized infusion of corn meal filtered through "porous earthenware vessels." In July, he performed inoculation tests on immature Bartlett pears with both his "cultured" bacteria and the bacterium-free filtrate. The pears injected with bacteria became blighted; those injected with the filtrate did not. Arthur plainly stated:[15]

> The evidence is thoroughly satisfactory and conclusive – the bacteria accompanying the disease of trees known as pear blight when fully isolated will produce the disease, while the juices in which they live will not. They are therefore the direct cause of the disease.

At the 1885 AAAS meeting in Ann Arbor, Michigan, Arthur presented his results as "absolute proof" of the bacterial origin of fire blight and confirmation of Burrill's earlier work.

Arthur primarily went beyond Burrill by moving from inoculating with exudates to using cultured bacteria grown in a variety of media. He found they grew quite vigorously on corn meal, hay, starch, and various vegetable substances. His attempt to create the blight by transferring other bacteria from rotting plants failed. When he mixed *Micrococcus amylovorus* with other bacteria and injected them into healthy trees, the resulting exudates from blighted tissue contained only *M. amylovorus*. These experiments further confirmed that one type of bacterium was responsible for fire blight. He also confirmed that the blight organisms in pear, apple, and quince were of the same species.[16] Although bacterial pathogenicity was still not yet accepted universally, in June of 1886, Arthur submitted his research on fire blight to Cornell University as a thesis and received his doctorate.[17]

There were two major mysteries involving fire blight that had not been solved by Burrill and which Arthur investigated. The first was how it progressed inside a tree and the second how it was transmitted between trees. Interestingly, Arthur noted a difference in the rapid motility of bacteria in culture and that of those pathogens taken from diseased trees, but did not account for it because he had not noticed the flagella present in culture.[18] Inside diseased trees, Arthur observed that the bacteria attacked the starch of cells first, then the cellulose of cell walls, and finally liquified the whole tissue. He observed that under other conditions, the cell walls were unaffected by the bacteria. Irrespective of the pathogen's effects on host cells, Arthur theorized that the movement of the bacteria throughout the tree "is doubtless largely due to simple displacement as multiplication takes place, although aided by the limited activity of the organism and the movement of the sap."[19]

Arthur was unable, however, to determine with certainty how the bacteria moved between trees. He smeared the "virus" over the outside of branches, leaves, and fruit. He tied diseased branches with healthy ones. He arranged an "apparatus . . . to draw air across diseased branches upon healthy ones." He even watered potted pear trees with water that was white with living, blight-causing bacteria. None of these procedures succeeded in transmitting the pathogen and causing subsequent development of blight.[20]

However, with more time and observation, Arthur began to form theories about transmission. He named three vulnerable points for blight bacteria to enter trees. After observing English hawthorn, Arthur suggested that the "germs" entered trees in the spring through the flowers and tender surfaces of young buds, recalling the discussions between Burrill and George Peffer. "The short period that the flowers are open," Arthur wrote, "is time enough to seal the doom of many limbs, and even whole trees."[21] The second entry point was the growing tips of branches or a developing bud, and the third was a crack or injury to the bark.[22] And almost as an afterthought Arthur allowed that insects might "now and then" be the agent of blight transferal.[23]

In 1886, Arthur discussed "preventatives and remedies" in an article for the U.S. Department of Agriculture. However, he could offer little more than Burrill had five years before. "No method is known," Arthur wrote, "or has yet suggested itself of rendering the tree insusceptible to the disease." Spraying had showed little success so far and Arthur noted that the internal surface of the flowers "are so well protected by the stamens and other organs that the antiseptic used does no service." Washes were another old method of delivering germicides, but they were also problematic. First of all, washes could not be applied to flowers or growing shoots, but "excellent results may reasonably be expected when made to the trunks and larger branches." Some growers had reported success with a traditional solution, i.e. sulfur mixed with lime. Further study needed to be made, but Arthur suggested using linseed oil for an elastic coating to which sulfur and 1% carbolic acid should be added.[24] The best method of decreasing the damage of fire blight remained cultural practices.

In 1887, Arthur left the New York Agricultural Experiment Station at Geneva for a position as a professor at Purdue University in Lafayette, Indiana, as well as the botanist at the Purdue University (Indiana) Agricultural Experiment Station. He remained at Purdue until his retirement in 1915. This transition marked the end of Arthur's association with fire blight. At Purdue, he shifted his science primarily to mycology and particularly the study of rust. He became a world-renowned expert in the field of Uredinales, so much so that his contributions to fire blight and phytobacteriology were largely obscured in the shadows of his own mycological accomplishments. Arthur himself, when writing a memoir for the Iowa State alumni records, mentioned his days at Geneva but highlighted his work with oat smut. He did not write a word about fire blight. By the time of Arthur's death in 1942, his biographers hurried past his role in the early fire blight work so they could discuss the vast wealth of his later research.[25]

There have been some recent attempts to reassess and compare the work of Burrill and Arthur. The distinguished plant pathologist K. F. Baker wrote a well-argued piece that sought to replace Burrill with Arthur as the man who should be credited with discovering that bacteria could cause plant disease. Baker believed that Burrill actually failed to prove his case, in a scientific sense, but rather merely suggested it as a likely possibility. Burrill's contributions were those of a pioneer certainly, Baker argued, but lacked scientific rigor and, most importantly, he failed to meet the standards of Koch's postulates. It mattered little to Baker that Koch's postulates had not yet been codified as the standard of proof when Burrill performed his experiments; the postulates merely represent a level of rigor that it is necessary to satisfy irrespective of whether they had been written down at the time. Rather, it was Arthur who met the test and demonstrated with experimental certainty that bacteria could cause plant diseases.[26]

Baker, however, is arguing against nearly a century of history. In the minds of most plant pathologists, Burrill has continued to hold the position of priority for the first recognition of the bacterial cause of fire blight. Burrill is the name plant pathologists associate with fire blight. Arthur, if his fire blight work is considered at all, is generally considered to have confirmed, refined, and expanded Burrill's observations. Still, there can be little doubt that Arthur's conclusions regarding fire blight were better founded than Burrill's and constituted stronger proof.

There are fewer examples of Arthur's work on fire blight from which to choose. Clearly, his piece to the AAAS presentation, "Proof That Bacteria are the Direct Cause of the Disease in Trees Known as Pear Blight," is his major contribution. In it, he both confirms Burrill's work and goes beyond it. Like Burrill, Arthur published this same information in a variety of places including the annual report of the Geneva experiment station and in the specialty magazine *Botanical Gazette*. The AAAS version of his research remains the best example of Arthur's approach to the subject. While he worked at an experiment station and, therefore, had practical, agricultural concerns too, he was clearly more focused than Burrill on the biology of the pathogen than on creating disease remedies.

The second paper by Arthur in this collection is "History and Biology of Pear Blight." This paper, as published in the Proceedings of the Philadelphia Academy of Natural Sciences, provided the most complete discussion and description of *E. amylovora* yet. In addition to presenting the specifics of Arthur's findings on fire blight, this paper is important for several other reasons. By being published through a natural history society as well as providing the content for Arthur's doctoral dissertation, it again highlights the dichotomy of plant pathology. In fact, the path of Arthur's early career is an excellent metaphor for the changing professional face of a science that has its roots in the amateur world of natural history and agricultural practitioners. The other great significance about "History and Biology of Pear Blight" which makes it worthy of reprinting is that it represents the first Ph.D. dissertation that involved plant disease studies granted by an American university.

The transition from Burrill to Arthur symbolizes a significant period for the science of plant pathology. It demonstrated a transition to a new level of professionalism and had a symbolic power that showed an advancing level of sophistication on the part of plant disease studies. Arthur was, in many ways, the first true American phytopathologist. While several major universities granted Burrill advanced honorary degrees, Arthur's doctorate marked him as a new generation of scientist, more of a 20th century biologist, separate from Burrill's tradition of the 19th century naturalist.

LITERATURE CITED

1. Kern 1942, 617.
2. Cummins 1978, 20-21.
3. Arthur [1878] 1981, 219-234.
4. Galloway 1900, 193-194.
5. Arthur, J. C. 1925. Botany and botanists in the eighties. In Biographical Files, National Fungus Collection; Trelease 1925, 61; Cummins 1978, 20.
6. Arthur, supra note 5, 8-12.
7. Thwaites 1900, 772; Arthur, supra note 5, 12-16.
8. Arthur, supra note 5, 13.
9. Ibid., 14.
10. Ibid., 14.
11. Rodgers 1952, 111.
12. Arthur 1885c, 295.
13. Arthur 1885a, 357-367.
14. Arthur 1885c, 296.
15. Arthur 1885c, 298.
16. Arthur 1885b, 1177-1185.
17. Rodgers 1952, 111; Kennedy, Widkin, and Baker 1979, 3.
18. Arthur 1886, 332.
19. Arthur 1885b, 1180-1181.
20. Arthur 1885b, 1183-1184.
21. Arthur 1887, 126.
22. Arthur 1887, 127.
23. Arthur 1885b, 1184.
24. Arthur 1887, 127-128.
25. Arthur 1916, 2; Kern 1942, 617; Mains 1942, 601-605; Cummins 1943, 328-331; Cummins 1978, 19-30.
26. Baker 1971, 603-633.

Fig. 3. J. C. Arthur, ca. 1886. (Courtesy of the National Archives.)

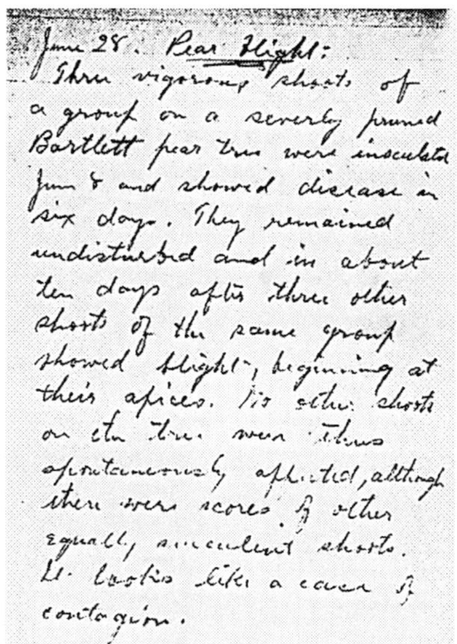

June 28. Pear Blight:
Three vigorous shoots of a group on a severely pruned Bartlett pear tree were inoculated June 8 and showed disease in six days. They remained undisturbed and in about ten days after three other shoots of the same group showed blight, beginning at their apices. No other shoots on the tree were thus spontaneously affected, although there were scores of other equally succulent shoots. It looks like a case of contagion.

Fig. 4. Excerpt from J. C. Arthur's laboratory notebooks on fire blight research, 1884-1886. (Courtesy of the New York State Agricultural Experiment Station, Geneva.)

Proof that Bacteria are the Direct Cause of the Disease in Trees Known as Pear Blight.*

J. C. Arthur

It has now been five years since Prof. Burrill brought the subject of pear blight before this Association and announced that it was due to bacteria. Previous to that time no instance of bacteria acting the rôle of vegetable parasites had been known, and the discovery was therefore a very important one, opening the way to a new and promising field of research.

The experiments of Professor Burrill showed that the disease alluded to was invariably accompanied by a specific form of bacteria (since named *Micrococcus amylovorus* Burrill), and that as the disease progressed a colorless or yellowish viscid substance was formed, apparently by the action of the bacteria upon the starch and other substances of the plant. The disease results in the complete death of all those parts of the tree that are attacked.

Although from these and subsequent investigations, the theory has been quite generally accepted that the bacteria are the cause of the disease, no rigid proof of it has yet been brought forward. It was with a view to either absolutely prove or disprove the theory that a course of experiments was begun last March and continued to the present time.

It had been incontestably shown that the disease may be readily transmitted to healthy tissues by introducing a drop of an infusion made by putting some thin slices of the diseased tissues in water,[1] or by simply transferring a minute portion of the exudation from the diseased to the healthy shoot.[2] The problem was consequently narrowed down to one of two alternatives, either the bacteria were the cause of the disease or the juices which accompanied them were the cause of it.

The first attempt was to secure some inoculating material in which the bacteria were entirely freed from the juices of the disease. This was done by

---

[1]Arthur, Bull. N.Y. Agric. Exper. Station, XCII: 3rd Ann. Rep. N.Y. Ag. Exper. Station, p. 358.

[2]Burrill, Amer. Assoc. Adv. Sci., XXIX, p. 589; Amer. Naturalist, XV, p. 529. Arthur, l.c.

*Originally published: 1886. *Proceedings of the American Association for the Advancement of Science*. Salem: AAAS. 295-298.

means of a succession of artificial cultures in a sterilized infusion of corn (maize) meal. Two series were successfully carried through, extending over about four months, and an inoculation from the sixth culture of each introduced into the green fruit of a Bartlett pear.

No. of Culture.	Date of Infection.	Kind of culture vessel.	Amount of fluid.
301	Mar. 27,	Salmon culture tube,	20 cc.
302	Apr. 1,	" " "	20 "
307	" 21,	Sternberg culture flask,	½ "
309	" 24,	" " "	½ "
314	May 22,	" " "	½ "
322	June 5,	Test-tube with Fol stopper,	25 "
364	July 13,	Bartlett pear on tree.	

The accompanying tables will show the kind of culture vessels used, the amount of culture fluid they contained, and the date at which each one was started. The first culture of the series was infected with a very small fragment of wood taken from the inner portion of a diseased limb of Flemish Beauty pear, in which the disease had been slowly advancing during the winter, from an inoculation made July 26, 1884, with an infusion of blighted twig from an apple tree. The infection of each of the other cultures of the series was successively made with a small drop of the one preceding. From the last culture a drop was transferred to a puncture in an unripe Bartlett pear. In both cases the pears were soon filled with the disease. They did not turn brown about the wound as is the more usual way, but first indicated the presence of the disease by beginning to shrivel, which occurred in one case, in nine days after inoculation, and in the other in ten days. Upon cutting the pears open the softer tissues were found broken down and liquefied, and a milky viscid juice ran out, showing that the disease had taken thorough possession.

No. of Culture.	Date of Infection.	Kind of culture vessel.	Amount of fluid.
301	Mar. 27,	Salmon culture tube	20 cc.
302	Apr. 1,	" " "	20 "
307	" 21,	Sternberg culture flask,	½ "
309	" 24,	" " "	½ "
310	" 27	" " "	½ "
320	June 5,	Test-tube with Fol stopper,	25 "
362	July 13,	Bartlett pear on tree.	

By this means of fractional culture the juices accompanying the bacteria first introduced were so much diluted in the transfers to succeeding cultures that the final drop used to inoculate the pear was practically free of them, and to the bacteria only, supplied by continued growth and multiplication, can we ascribe the last result.

Having shown that the bacteria when isolated from their juices are able to cause the disease, it still remained to show what action the juices would have when separated from the bacteria. On July 18, a strong infusion of blighted pear was filtered through a porous earthenware vessel, such as used for small electric batteries, and an unripe Bartlett pear inoculated with the filtrate and another pear with the infusion used for filtering. In a week the latter was thoroughly blighted, while the former showed no signs of injury except the slight wound, which finally healed. On July 24, another strong infusion of pear blight was filtered through a second battery cell. Both of these cells were new, and had never before been used for any purpose. In this case two unripe Bartlett pears were inoculated with the unfiltered infusion and two others on the same tree with the resulting filtrate. Both the former showed strong evidences of disease within three days, while the latter soon healed up the small wounds made by the inoculation, and have continued their normal growth.

The evidence is thoroughly satisfactory and conclusive – the bacteria accompanying the disease of trees known as pear blight when fully isolated will produce the disease, while the juices in which they live will not. They are, therefore, the direct cause of the disease.[3]

---

[3]This paper is also published in the Botanical Gazette, vol. X, p. 343, and Gardeners' Chronicle, vol. XXIV, p. 586.

History and Biology of Pear Blight.*

J. C. Arthur

To the American orchardist or nurseryman the name of pear blight, or fire blight, as it is often called, brings to mind a serious malady of fruit trees, which has been the theme of incessant discussion by horticultural writers and speakers since the earliest days of fruit culture in this country. The most marked features of the disease were admirably characterized by William Coxe[1] at the beginning of the present century, in the following words: "That species of blight which is sometimes called the fire blight, frequently destroys trees in the fullest apparent vigour and health, in a few hours, turning the leaves suddenly brown, as if they had passed through a hot flame, and causing a morbid matter to exude from the pores of the bark, of a black ferruginous appearance; this happens through the whole course of the warm season, more frequently in weather both hot and moist." The disease occurs from Canada and Minnesota on the north, to Georgia and Louisiana on the south, and from the eastern limit of the Rocky Mountains to the Atlantic ocean (sic). No part of this vast extent of country is exempt, although it does not appear with the same frequency and power in all localities, and is usually rare in the immediate vicinity of the sea-coast.

So far as at present known, it is exclusively confined to this part of North America. This is partly inferred from the absence of any distinct mention of such a disease in the horticultural literature of other regions, and partly from direct testimony. Prof. Dwinelle, late of the University of California, has told the writer that it does not occur on the Pacific coast. Dr. De Bary[2] whose word carries great weight, says, after giving a brief description of the disease, "this phenomenon is not to my knowledge known in Europe." A long account of the disease has been published by Dr. Wakker,[3] in a gardening journal of Holland, in order to learn if it occurs in that country, but up to the

---

[1] Cultivation of Fruit Trees, Philadelphia, 1817, p. 174.

[2] Vorlesungen über Bacterien, 1885, p. 137.

[3] Nederlandsche Tuinbouwblad, II (Jan. 9, 1886), p. 9.

*Originally published: 1886. *Proceedings of the Philadelphia Academy of Natural Sciences.* 322-341.

present time no one has intimated any knowledge of it. In a recent letter Dr. Masters, editor of the Gardeners' Chronicle of England, says that no such disease has been recognized in the British Isles. The testimony of one of our own horticulturists, Prof. Budd,[4] of Iowa, who is familiar with the disease in this country, and has inspected the orchards of the old world far into Russia, is especially valuable; he says "no trace of blight of pear- or apple-trees can be seen in Europe." From these statements, and the inferences to be drawn from other sources, it appears highly probable that the disease does not extend to Europe. An account of the principal diseases of fruit-trees of New Zealand, by Prof. T. Kirk,[5] has been published, which describes a disease of the pear known in that country as *fire blight*, due to a fungus, and another of the apple, the *American blight*, due to an insect. No true pear blight, as recognized in the United States, is mentioned, and in a recent communication the author has definitely stated that it is not known in the colony. Whether it occurs in other parts of the world is not yet ascertained, if some slight testimony regarding its absence in Japan be excepted.

It is only within a year or so that European writers have become aware of its existence, and this only through American authors. It is remarkable that a disease of such virulence and so easily transported should not have found its way across the ocean, when one remembers the number of destructive plant maladies that America has already involuntarily foisted upon European cultivators. It will not be profitable to speculate much at this time upon the reasons for this, but we may suppose that the small exportation of American fruit-trees, or of scions,[6] has been a factor in keeping it in check. The influence of climate, and some less evident factors, need not be discussed in this connection.

Amount of Loss. – It has already been intimated that pear blight is a frequent and destructive affection; it will tend to give a fairer appreciation of

[4]Trans. Minn. Hort. Soc. for 1883, p. 281.

[5]Fruit Blights and Diseases of Fruit-trees in New Zealand, 1885.

[6]For an account of the destruction of stored scions by blight, see Rep. Hort. Soc. of Mich. for 1881, p. 137.

the subject if it be stated how frequent and how destructive it is. Coxe[7] as early as 1817, in the oldest pomological work by an American author, says it "frequently destroys trees in the fullest apparent vigour and health, in a few hours; I have in twenty years lost upwards of fifty trees." The years 1826 and 1832 were notable in horticultural circles for the increased prevalence of the disease; but it was in 1844 that the most widespread and fatal epidemic, that the country has yet known, occurred. Few, if any, pear-orchards escaped at that time without the partial or total loss of many trees, and some orchards, even large ones, were quite destroyed. The following year the epidemic was much lighter, and had fully disappeared by 1846. Although it had subsided as an epidemic, it still occurred in localities here and there, and has continued to do so until the present time. Judging from the communications in the horticultural press, the whole country, or various sections of it independently, have been subject at various times to epidemic visitations, but none have equaled in severity that of the memorable year of 1844.

It is often maintained that a certain periodicity of occurrence is observable, the periods usually being placed at five, ten, or twenty years. A careful examination of the literature of the subject, however, gives little support to these views, and makes it more probable that the intervals are irregular, and that they vary for different sections of the country. The year of maximum prevalence may or may not be preceded by one in which the disease is noticeably common, but it is quite invariably followed by a year or two of successive decadence.

In the absence of exact statistics, which it has not been practicable to obtain, something of the important nature of the disease may be gathered from the statements of horticultural writers and the phraseology which they employ in speaking of it.

The renowned horticulturist, A. J. Downing,[8] called it the "monstrous malady of the pear." Chas. R. Baker[9] says it is "the worst malady with which the cultivator of the pear-tree has to contend." In southern Pennsylvania "the

---

[7] L.c., p. 174.

[8] Horticulturist, vol. i, 1846, p. 62.

[9] Practical and Scientific Fruit Culture, Boston, 1866, p. 476.

pear is so generally destroyed by the blight," according to J. B. Garber[10] writing in 1850, "that very few trees are to be found." At Philadelphia, however, the disease has been rarely observed, according to Thomas Meehan.[11] T. T. Lyon,[12] of Michigan, states as the opinion of many cultivators in that State, that the pear-tree cannot be grown with financial success on account of the blight. Illinois has always been much subject to the disease, and Prof. J. B. Turner,[13] in 1868, gave expression to the general feeling of his region by describing it as "that deadly Upas of the pear-tree known par excellence as the pear-blight." In 1882 Dr. J. L. Hallam[14] speaking for southern Illinois, says, "pears have failed, utterly failed, so that none are now cultivated for market, the blight has destroyed the trees – branch and root," and S. G. Minkler,[15] in the northern part of the State, observes that it is a very uncommon thing to see pear-trees without dead branches or other signs of the ravages of blight. Wm. A. Nourse,[16] of the same State, is led to "doubt if one-tenth of the pear-trees that are set, live ten years," on account of this one destructive agent. Geo. M. Dewey,[17] of Missouri, says that "with good cultivation and rich soil the pear generally dies of blight before the eighth year." In Minnesota the severe climate has not permitted the cultivation pears, and almost the only apples grown for many years were the hardy crab-apples. The latter have been rapidly improved, and together with the hardier varieties of the common apple would now furnish this part of the country with an abundant supply of fruit, were it not for this same disease, which elsewhere most conspicuously preys upon the pear-tree. E. H. S. Dartt[18] held the opinion in 1874 that the severity of winter was not so much

---

[10]U.S. Patent Office Report for 1850, pt. ii, p. 418.

[11]Rep. Penn. Fruit-Grower's Soc. for 1877, p. 77.

[12]Rep. Pomol. Soc. of Mich. for 1878, p. 368.

[13]Trans. Ill. Hort. Society for 1868, p. 42.

[14]Same for 1882, p. 118.

[15]Same for 1882, p. 30.

[16]Same for 1880, p. 63.

[17]Proc. Mo. Hort. Soc. for 1870, p. 18.

[18]Trans. Minn. Hort. Soc. for 1874, p. 22.

to be dreaded as the ravages of blight. He had at that time one or two thousand trees affected. Dr. P. A. Jewell,[19] up to 1876, had lost one thousand Tetofsky apple-trees by it. F. G. Gould[20] says that "only for this scourge every family living on a farm in Minnesota could have a supply of apples."

Citations enough have doubtless been given, although several pages of equally strong ones might be added, to show that fruit growers, who have the best opportunities for observation, consider it a disease greatly to be dreaded and one of special economic importance. Other sections of the country, notably those of Ohio, western New York and Georgia, could furnish equally important proof of these propositions. All that is desired in this connection, however, is to give those not familiar with the subject some idea of the disease and its effects as ordinarily observed.

Early Records. – The oldest mention of the disease, that gives a good and reasonably full description of it, is in Coxe's work on fruit-trees, bearing the date of 1817. The manner of the author leaves no doubt that it was well known at that time, and the reference to his losses during twenty years makes it reasonably certain that he had observed the disease as early as the opening of the century. The earliest notice, however, which has yet come to hand, is in a letter written by Wm. Denning,[21] describing the disease in apples, pears, and quinces. He speaks of first observing it on the Highlands of the Hudson in 1780.

There is no interest, however, in tracing chronologically the various notices found in different publications, for without exception they have the tone of treating a familiar theme, and show no evidence that the disease in the first part of the century was in any respect different from to-day.

Conjectures Regarding its Cause. – A brief treatment of this topic will be all that is required for the purposes of this paper; and only those hypotheses will be touched upon which received such careful presentation as to attract the favorable attention of the public.

---

[19] Same for 1876, p. 73.

[20] Same for 1884, p. 127.

[21] Trans. Soc. for Prom. of Agric., pt. ii, 1794, p. 219.

Few writers appear to ascribe the disease to a single agency, but regard it as resulting from several causes, either acting together or brought about by dissimilar circumstances. Little discrimination is made between predisposing conditions and active agents. In fact sharply defined treatment could not be expected when all was conjecture, and when the shrewdest observers did not hesitate to avow that after years of loss under all kinds of experimentation, and after interminable discussions, the cause still lay shrouded in impenetrable obscurity.

Coxe,[22] who has had many followers, thought that the hot rays of the sun when acting through a misty or saturated atmosphere deranged the vital activities of the plant and brought about the disease. He considered old varieties more subject to it, on account of having lower constitutional vigor, than new varieties, of which the St. Germain and Seckel were respective ly conspicuous instances.

The insect theory, as it was called, was promulgated at this time. It was started upon firm facts by the discovery of a small brown beetle, about two millimeters long, which penetrated the branch, and caused the part beyond to die. The beetle received the name of *Scolytus pyri* Peck, now changed to *Xyleborus pyri* (Pk), and is still known as the blight beetle. The effect of its attack appears to the casual observer similar to that of the true blight – the branch in June or July rapidly withers, and the leaves and fruit turn black. The beetles being minute and inconspicuous escape attention, and the fact that the branch does not die below a definite point is sometimes overlooked. It is not difficult to see how many persons came to connect this comparatively rare affection with the common fire blight, and to believe that insects of some sort were to be held accountable for all – their supposed minuteness and wary habits being sufficient reasons for the failure to find them, and the spread of the disease along the limbs of a tree being ascribed to a poison which the insects were supposed to emit. Among the prominent supporters of this view was the "Genesee Farmer,"[23] published at Rochester, N. Y., with Patrick

---

[22] L.c., p. 175.

[23] See Genesee Farmer, vol. vii, 1846, p. 217; vol. viii, 1847, pp. 122, 218, etc.

Barry as the horticultural editor. It has not, however, been so strongly advocated for the last decade or two.

The next hypothesis that attracted general attention was known as the frozen-sap theory. This was based upon the supposition that the autumn or winter freezing of unripe wood produced a poison which the moving currents of sap the next spring and summer distributed, causing the death of the parts. It was first published in 1844 by Rev. H. W. Beecher,[24] of Indiana, in a long and able article in "Hovey's Magazine." In the following year it was independently elaborated by A. J. Downing[25] in his work on Fruits and Fruit-trees of America, who first called it the frozen-sap theory, and who is usually spoken of as the author of it. This view has probably had more firm adherents than any other, as it explained many phenomena connected with the disease in a fairly satisfactory manner. It was especially well received in the western States.

The next hypothesis which gained the attention of the public was the fungus theory. Its first successful presentation was in 1863 by Dr. J. H. Salisbury,[26] who figured the fungus which he decided to be the specific cause of this kind of blight, and ventured to give it a name, although he was sadly in error in most that he did. Thomas Meehan,[27] editor of the "Gardener's Monthly," has ably championed this explanation, and done much to keep it in favor. In 1875, Dr. J. G. Hunt,[28] by Mr. Meehan's request, undertook a microscopical examination of blighted pear-twigs, and confirmed the opinion that it was due to a fungus, without, however, deciding upon the specific character of it.

Blighted trees often attract attention immediately after thunder storm, and from this and other circumstances the belief that the malady is due to

---

[24] Magazine of Horticulture, vol. x, p. 441.

[25] Fruits and Fruit-trees of America, p. 324; same, 2d Revision by Chas. Downing, p. 646.

[26] Ohio Agric. Rep. for 1863, p. 450.

[27] Proc. Amer. Pomol. Soc. for 1867, p. 59; and elsewhere.

[28] Gardener's Monthly, vol. xvii, 1875, p. 245.

electricity has gained many adherents, but the argument has not had a full and connected presentation.

The last hypothesis of historical importance is the bacterial theory. Although hinted at by a number of horticultural writers, yet the credit of it is due to Prof. T. J. Burrill,[29] who in 1878 distinctly stated his belief that the cause resides with the bacteria which he found in great abundance in the tissues of affected branches. In 1880 he performed a series of experiments[30] by inoculating healthy branches with the juices of diseased ones, the results of which were presented to the American Association at its Boston meeting, thus first bringing the subject clearly to the attention of the scientific world. Although this was now the popular hypothesis, it cannot be said to have received more substantial credence than those which had gone before, either from the horticulturists or the scientists. The experimental results gained by Prof. Burrill were confirmed and extended by the writer[31] during 1884, by means of a similar series of inoculations.

Of the multitude of minor hypotheses which were put forth in explanation of phenomena connected with pear blight, and which were variously received, and of all degrees of plausibility, it is impossible to speak at this time without carrying this paper to undue length.

Beginning of Experimental Research. – The question of the cause of pear-blight was finally removed from the domain of speculation to that of established fact by a series of crucial experiments performed by the writer[32] a year ago, and recorded in a paper before the American Association at the Ann Arbor meeting. These consisted in showing that the bacteria when removed from the tree and passed through a series of artificial cultures would generate the disease when again introduced into the tree, and that the juices

---

[29]Trans. Ill. Hort. Soc. for 1878, p. 80.

[30]Proc. Amer. Assoc. Adv. Sci., vol. xxix, 1880, p. 583; Rep. of Ill. Industrial Univ. for 1880, p. 62; Trans. Ill. Hort. Soc. for 1880, p. 157; Amer. Naturalist, vol. xv, 1881, p. 527.

[31]Rep. N.Y. Agric. Exper. Station for 1884, p. 357.

[32]Proc. Am. Assoc. Adv. Sci., vol. xxxiv, 1885, p. 295; Bot. Gazette, vol. x, 1885, p. 343; Gardeners' Chronicle, vol. xxiv, 1885, p. 586.

accompanying blight when cleared of bacteria by filtration will not produce the disease.

Having now come to a firm basis for scientific advancement, let us look over the historical ground again to see if some one did not hit upon the true explanation of the disease, although he may not have been aware of its significance. In a connection like this, facts derived from experiment have greater weight than statements of opinion; the latter acquire importance in proportion as they are logically derived from correct and close observation. Bearing this in mind, we need not give much heed to the not uncommon inference that pear-blight was in some intimately related to the epidemic diseases of man, *e.g.* cholera. This view is said to have been quite frequently entertained in the early part of the century, but was not sanctioned by the learned. The use of such phrases as "first cousin to the cholera," "a species of vegetable ferment," etc., surely does not entitle the author to any priority in way of discovery.

We turn from these slight hints to the record of an experiment in inoculation, made in 1845 and published the following year. We are told by S. B. Gookins,[33] of Indiana, that visiting Mr. Ragan (the same person who furnished Mr. Beecher with many of the facts on which he founded the theory to which we have (already referred) he was shown a thrifty young pear-tree in the nursery, which had been "inoculated" "by way of experiment" with "the sap of a blighted tree," "a few days previous." "He made an incision about three feet from the ground, lifted the bark as in the process of budding, and injected a small quantity of the diseased sap." "We found the leaves of the patient changing color, and emitting that peculiar odor which is always present in cases of blight, and upon applying the knife, the inner bark was found to be black from the root to the top, while nothing of the kind appeared elsewhere in the nursery."

This admirable experiment was combined with a no less admirable interpretation of the cause of blight. The writer cites facts to disprove the hypothesis of Mr. Beecher, and then says: "I strongly incline to the belief, that the pear-blight is an epidemic, that it prevails like other epidemics, and

---

[33] Horticulturist, vol. i, 1846, p. 253.

will pass off like them. The atmosphere is, I believe, generally admitted to be the medium by which they prevail, and are carried from place to place. What that subtle principle may be, which pervades our atmosphere, by which infection is retained and transmitted, human science has not discovered; but that such a principle exists is sufficiently obvious from its effects."

This clearly conceived elucidation of the matter could only have been improved by a knowledge of the germ theory of disease, and when we remember the date at which it was uttered, we do not feel that the writer was guilty of any lack of acuteness in not perceiving the relation which we now know to exist between his theory and his facts. He seems to have been a modest man, for he only signs his initials and does not defend his views when the editor, A. J. Downing,[34] opposes the opinion that "an epidemic conveyed by the atmosphere is too slightly supported by facts to weigh at all against the observations of cultivators," which "strongly point to the freezing of the sap as the cause."

Another interesting experiment was performed by Dr. E. S. Hull,[35] of Illinois, in 1870. Having received some blighted apple twigs from a correspondent, he cut pieces from them with which he "inoculated several succulent pear shoots by tying in the pieces as in budding." This was done the middle of June, and no observation taken for thirty-four days, when the blight was found to have extended several inches into the healthy tissues. From this he very justly concludes that the blight in apples and pears is but one disease, but seems to take it for granted that in both it is due to "vitiated sap."

The fact that the disease may be transferred to healthy trees by the pruning knife has been observed by several persons. H. Wendell,[36] of New York, says in 1849, "I am also careful that the blade of the knife is perfectly clean, and that it has none of the sap of a diseased tree adhering to it, because I have known many valuable trees destroyed by having been inoculated in

[34] Same, p. 255.

[35] Trans. Ill. Hort. Soc. for 1870, p. 220.

[36] U.S. Patent Office Rep. for 1849, pt. ii, p. 447.

this manner." Prof. Turner,[37] of Illinois, makes a similar statement: "I found that this disease is exceedingly contagious, for if I used my knife to prune a healthy tree after having used it in shaving the diseased one, I communicated the disease to that tree."

Prof. Burrill first observed the bacteria of blight in 1877,[38] but did not recognize them as such till the following year,[39] when he avowed his belief that they were the cause of the disease. His first inoculation experiments were made in 1880, as already stated. In 1882 he characterized the organism under the name of *Micrococcus amylovorus.*[40]

Description of Micrococcus amylovorus Bur. – The form of this species of bacteria is very constant, under all conditions. The single cells are from oval to roundish-ovoid, and only vary by slight changes in the ratio between their length and breadth (Pl. III, figs. 1, 2, 6). They are 1 to 1¼ µ long, by ½ to ¾ µ broad, and quite colorless. For the most part, they exist as single independent cells, but may often be found in pairs, especially when still multiplying, and in rare instances are united into a series of four or even more, but never extend into chains.

During rapid vegetation, in rich nutritive media, the movements reach a stage of extreme activity. The appearance is what is termed swarming, in which the bacteria move rapidly back and forth, in and out among each other, but never in a straight line to any distance. As the rate of growth becomes less from any cause the movements are retarded. Taken directly from the tissues of a blighting tree, the movements of translation are usually sluggish or imperceptible, although the universal Brownian movement is likely to give a misleading appearance of activity. Under specially favorable conditions, as when grown during hot weather in very succulent shoots, or

---

[37]Trans. Ill. Hort. Soc. for 1878, p. 81.

[38]Same for 1877, page 114.

[39]Same for 1878, page 79.

[40]The Bacteria (a reprint from Rep. of Ill. Industrial Univ. for 1882), p. 42; Amer. Naturalist, vol. vii, 1883, p. 319. In the last publication, by a typographical error, the name was made to read *M. amylivorus,* a mistake which has been copied into other works – see Grove's Bacteria and Yeast Fungi, London, 1884, p. 10.

from artificial inoculation in unripe fruit, the movements are much increased and may become quite rapid. When taken from the tree in winter, or when grown in solutions that are too acid or too alkaline, or which are deficient in the proper nutritive substances, there is no perceptible locomotion.

When in active growth, the cells present a uniformly dull appearance. By conditions which are unfavorable to normal growth, yet do not entirely check it, such as strongly acid or alkaline solutions, deficient nutriment, or exhaustion by keeping the cultures several months, the cells become highly refractive, and to some extent take on the appearance of the spores of other species of bacteria. Whether in this state they possess any of the characteristic powers of resistance which belong to spores, has not been ascertained.

Formation of Zooglæa. – By far the most characteristic feature in the life history of *Micrococcus amylovorus* is the formation of zooglæa (figs. 2, 3, 5). These have never been observed in the tissues of the tree under any conditions, or in or upon any sort of solid media, but they occur with much regularity in fluid cultures, when placed under favorable conditions for rapid growth.

They are produced to some extent throughout the fluid, but are most abundant in the thin pellicle which forms upon the surface, appearing within forty-eight hours from the beginning of the culture. The substance of the pellicle consists of a colorless matrix uniformly filled with motionless bacteria, and against this the zooglæa are sharply defined. They are often brought out yet more distinctly by being surrounded by a colorless layer, free of bacteria, which is doubtless an extension of the ground substance of the zooglæa mass (fig. 4).

The masses are far more dense than the pellicle, and are compactly filled with refractive bacteria. They possess a definite outline, and are recognizable when very small; and although they may reach 30 to 40 μ long by 20 to 30 μ wide, they rarely lose their distinctness. When below 10 μ in length, their usual form is oblong, varying to globular. They occur singly, or united more or less intimately end to end in pairs, and sometimes several form a short chain. At this stage they possess a uniformly even and unbroken surface, which now becomes uneven and wrinkled and is finally thrown into folds, giving some resemblance to the external aspect of the brain. Zooglæa more than 20 µ in length have the folds of somewhat unequal height, and the sinuses deeper giving a strong cerebric look, or when the folds are small and circular, they are better described as mulberry-like. The elongated forms, which at some stage of growth might doubtless have been composed of two or more distinct masses, often take on a vermiform appearance. But whatever the variations may be, the distinctness of outline, the general form, and the cerebric surface are unfailing characters, which so far as my knowledge extends, are not found in any other species of bacteria.

Cultivation in Fluid Media. – The range of substances which may serve as culture media for this organism is very wide. An infusion of almost any vegetable substance containing a fair amount of soluble carbohydrates is likely to be sufficient to enable growth to take place, even if not very luxuriantly.

The substance which on the whole has proved most satisfactory is an infusion of potato. This is prepared by paring a potato and slicing it into three or four times its bulk of water. This is kept for a couple of hours at about 70° C., by placing it over a water-bath, during which time it is occasionally stirred. It is then filtered, and is ready to be placed in the culture vessels for sterilizing and use. If the heat is allowed to rise much above 70°, the starch is gelatinized, and it is only with difficulty that the solution can be filtered. The resulting liquid is clear and watery, but is often light brown from coloring matter contained in the potato, which does not, however, materially interfere with observation on the growth of bacteria in it. Iodine gives a blue coloration to this liquid, showing that it contains starch, probably in the form of amylon. Another equally good culture fluid is made by treating corn (maize) meal in a similar manner. The solution is colorless, but it is very apt to throw down a troublesome sediment, which makes it less desirable to use than the potato solution.

Test-tube or flask cultures with these liquids, when kept at a temperature of 25° to 30°C., usually show some turbidity in twenty-four hours after being infected, and if the growth is very rapid, bubbles of gas (CO_2) will be given off, which collect at the surface into a slight froth. In forty-eight hours the liquid has become thoroughly turbid. By this time a thin whitish pellicle has formed on the surface, which does not increase much in thickness up to the end of active growth, and rarely becomes wrinkled. With the formation of the pellicle, a sediment gathers at the bottom of the liquid, often a centimeter in depth, but which is so light that it only apparently differs from the liquid above by being whiter. In the course of some weeks this sediment will mostly gather upon the bottom of the vessel. No difference has been observed in the appearance of the bacteria taken from different parts of the culture. Those imbedded in the pellicle are not arranged in any recognizable order.

In proportion as liquids are less suitable to the growth of the organism, the visible changes are less. The pellicle may not be formed, and there may be no turbidity, but if any growth at all takes place there will be some evidence of it by formation of a slight sediment. But the occurrence of a precipitate does not necessarily imply growth, for it not infrequently separates from a liquid containing organic matter, although remaining perfectly sterile.

An infusion of hay, and also of dead, partly decomposed grass from a marsh, gave nearly a normal growth of blight bacteria, but the cells were considerably more refractive than usual.

A solution of starch, having one part of starch to fifty of water, gave but a slight growth of highly refractive bacteria, without a pellicle, turbidity, or zooglæa. A strong decoction of old barnyard manure acted in the same manner. A solution of one part of glucose to fifty of water gave no growth of the bacteria.

In testing the effect of acids upon the development of blight bacteria, a ½ per cent. of malic acid was added to the usual infusion of potato. This prevented the formation of a pellicle, turbidity or zooglæa, but gave a very considerable cloudy sediment, largely made up of loosely aggregated groups of blight bacteria, which were brilliantly refractive. A similar solution with 2 per cent. of malic acid gave a slightly less abundant sediment, but with otherwise the same results. Some of the latter was transferred to a corn-meal solution, producing the characteristics of a pear blight culture, except the formation of zooglæa. After some days this was introduced into a pear tree, which in due course of time gave the true blight, showing that the bacteria of the acid solution were really blight bacteria. Attempts to grow them in a nutrient 5 per cent. solution of citric and tartaric acids have not been successful.

Testing the nature of the bacteria in cultures producing limited growth, by inoculating directly into the tree, has not, as a rule, proved successful, as for some reason they seem unable to gain a footing in the living tissues. It is therefore necessary to transfer them first to richly nutrient cultures, from which, after a time, they may be introduced into the tree, and, if the blight bacteria are present, will start the disease.

Cultivation in Solid Media. – In test-tube cultures with nutrient gelatine the most characteristic results have been obtained by adding a drop containing blight bacteria to the gelatine while liquid, and thoroughly distributing the germs by shaking the tube. In from two to three days the gelatine contains numerous small white dots, which, upon examination under the microscope, prove to be a mass of bacteria of the usual appearance. The dots are globular or oval, and increase to about .5 mm. in diameter. No further growth or change takes place, and in this condition they remain for weeks, without liquefying or otherwise affecting the gelatine.

When sown upon the surface of the gelatine by drawing a needle or glass rod over it, or by placing a drop on it, the growth is feeble and does not amount to more than a slight shining, appearance of the surface.

A nutrient solution made from an unripe pear, in which blight germs were well distributed by shaking and then left undisturbed for two days, gave the same isolated white dots as in gelatine; but they dropped to the bottom of the liquid upon being jarred. The bacteria were evidently prevented from moving freely by the jelly, which was not, however, thick enough to keep the groups in place when its cohesion was once disturbed. Fruit jellies, doubtless, may be found to be convenient media for the cultivation of this species of bacteria.

No success has been attained in the use of agar agar, but whether due to a want of adaptability in the substance, or to wrong manipulation, must be left to future experiments to determine.

The opaque solid cultures proving most successful have been conducted upon freshly gathered unripe pears. Slices of these are placed under a moist bell-jar, and infected by touching with a needle that has been dipped in some substance containing the bacteria. In two or three days fine milky drops, like beads of dew, will appear scattered over the surface for 5 mm. or more about the infected spot. These will become somewhat larger after a time, while the spot which received the infection will turn slightly brown, the tissues gradually wasting away and forming a small depression.

If, however, the slices, having freshly cut surfaces both above and below, are laid upon a plate with a little water, and placed under a bell-jar, the result is not the same. The dew-like drops appear within forty-eight hours, as in the other case, but increase rapidly in size, while a drop is also formed at the point of infection. Drops finally appear over the whole surface of the slice. They remain more or less distinct, and soon become as large as a pea, retaining the globular or rounded form to a remarkable degree. Microscopically they are composed of the usual form of blight bacteria, suspended in a colorless fluid. After about a week, the drops coalesce and the tissues of the pear begin to break down. This sort of culture requires no precautions of sterilizing, as no other bacteria can multiply upon it till after the cells of the pear begin to die.

When blight bacteria are sown upon slices of baked or boiled potato, they spread out over the surface in a thin, slightly moist layer, which is usually somewhat yellowish, but do not grow readily, or produce a characteristic appearance. Under the microscope the cells are strongly refractive.

A boiled potato was infected by thrusting a platinum wire, smeared with blight bacteria, into one end. After sixteen days it was cut open. No external change had taken place, and, to the unaided eye, no internal change either; the odor and texture were still those of a freshly boiled potato. The microscope, however, revealed the blight bacteria in every part of the potato, in irregular motionless masses, and with more than the usual refractiveness.

These opaque solid cultures have brought out one fact very distinctly, which is, that *Micrococcus amylovorus* requires a large supply of water for its best development – a fact which has an economic bearing.

Behavior toward Staining Fluids. – So far as trial has been made, nothing especially characteristic has been detected to distinguish this form of bacteria from the majority of micrococci. The most successful results have been obtained with a watery solution of Bismarck-brown, especially in cover-glass preparations. These make excellent specimens when mounted in Canada balsam.

The zooglæa are inclined to be too deeply stained by this process, and for most purposes they are best studied unstained. They may be well preserved by mounting in glycerine.

Hæmatoxylin has also given good results, but has not been found particularly useful.

Chemical Products. – The chemical changes brought about by the activity of the blight bacteria have not yet been fully and carefully worked out. The most obvious product is carbon dioxide, which often passes off so freely from a cultivation as to produce a slow effervescence. Butyric acid and alcohol are formed in very small quantities, if at all. The tests by which these facts have been determined have already been published,[41] and need not be repeated here. Vigorous cultures of the bacteria in infusion of potato give no reaction for glucose with Fehling's solution; and blighting tissues from the tree give no indication by the same test of more than the normal amount of glucose to be found in healthy tissues. On the other hand a quantitative determination of sound and blighting pears, taken from the tree at the same time, shows considerable less sugar in the latter.

A favorite explanation with horticulturists of the action of fire-blight upon the pear-tree, has been to say that the sap is poisoned. This poison was supposed to be introduced by insects, or to be due to some disorganization of the tissues. Although it is now known that specific bacteria are directly answerable for the disease, it is yet worth while to see if the old idea of a poison has not some foundation in fact.

It has been ascertained that certain bacteria produce during their growth, characteristic poisons which are classed under name of ptomaines. Most of the ptomaines are non-volatile, and readily soluble in water or alcohol. The chemical tests which are applied for their detection cannot be considered conclusive except when taken collectively. The tests tried below are among the most satisfactory known at present.[42]

A cultivation in infusion of potato, giving about 200 cc. of liquid was filtered, and the filtrate evaporated to a syrup. This treated with alcohol, and

---

[41] Rep. N.Y. Agric. Exper. Station for 1885, p. 247; and less fully in Amer. Nat., vol. xix, 1885, p. 1181.

[42] Cf. Brieger, Ueber Ptomaine, 1885, p. 22, *et seq.*

the solution tested with the most characteristic test for ptomaines – the reduction of potassic ferriccyanide. Other portions of the solution were successively tested with phospho-molybdic acid, potassio-mercuric iodide, and iodine in potassic iodide, all of which failed to give any distinctive reactions.

Another trial was made with about 200 cc. of material prepared by cooking a potato in just enough water to cover it, sterilizing, and cultivating the bacteria in it as usual. In four days from beginning of the culture it was filtered; the residue upon the filter was treated with 100 cc. of distilled water, slightly acidulated with hydrochloric acid, heated to 70°C. and filtered. The two filtrates were united and evaporated to a syrup. This was digested in the cold with alcohol containing a little sodic hydrate. This solution was tested as before, and also with platinic chloride and concentrated sulphuric acid, and all with no distinctive reactions.

A third trial was made with a boiled potato, which had been permeated with the blight. The extract was made by the Stas-Otto method, and the same reagents used as in the last case, with equally negative results.

These tests do not cover the possibility or the ptomaine being volatile, which is really not very great. It is yet necessary to make tests of freshly blighted tissues from the tree, which can only be done during the hot months.

Action of the Organism in the Living Plant. – The bacteria of blight have the power of growth and multiplication in the presence of the living cells of the pear, and in this one important respect differ essentially from other species of bacteria. By artificial inoculation into growing unripe pears, which give most marked and certain results, it is found that other bacteria are entirely innocuous, at once disappearing without having made any growth or induced any changes in the tissue of the pear. If blight bacteria in active condition are intermixed with the other forms, they penetrate the cell, multiply, and finally bring about the disorganization and death of the tissues which marks the progress of the disease, but the associated forms disappear the same as when introduced alone, and the product is a mass of practically pure blight bacteria.

This result is rendered possible on account of the fact already stated, that the blight bacteria penetrate the tissues, and maintain their normal growth for some time (days or weeks), before the life of the cells is sufficiently interfered with to permit growth of other forms. The bacteria always extend beyond the visible location of the disease – in small branches, often to the distance of a third of a meter or more.

One of the properties which enables this species to successful penetrate the pear-tree is evidently its unusual indifference to acids, which prevents most other forms from making any growth; the juices of the pear give a strong acid reaction with test paper.

What chemical changes are brought about by its activity in the plant cannot be definitely stated, further than to say that a mucilage or gum, which is soluble in water, is produced in abundance, with the disengagement of carbon dioxide. The contents of the cells, together with the cell-walls which have not been liquefied or changed into stony tissue, pass over into this viscous product.[43]

It was clearly observed by cultivators, being recorded by Coxe, that succulent shoots blight the most readily, and any process of cultivation which as far as possible prevents succulency has always been considered an aid in keeping the disease in check. The avidity of the blight bacteria for water has been well demonstrated in the cultures on slices of pears. There seems to be some connection between these facts and the well known fact that the disease shows different degrees of virulence in different varieties of fruit trees, especially in different varieties of the pear. The variation, or at least part of it, to be observed in pears, apples, quinces, hawthorns, etc., may be due to some inherent difference in the nature of the host, not readily formulated; for we find that the blight bacteria will grow to only a slight extent in the succulent peach shoots, and not at all in most other plants.[44] But in varieties of the same fruit it may reasonably be inferred that to a considerable extent the difference in the progress of the disease is due to physical causes.

---

[43]Rep. N.Y. Agric. Exper. Station for 1885, p. 248; Amer. Nat., vol. xix, 1885, p. 1181.

[44]Rep. N.Y. Agric. Exper. Station for 1884, pp. 362, 377; Amer. Nat., vol. xix, 1885, p. 1182.

To determine what relation the hydration of the tissues holds to this question, a series of determinations of the percentage of water in the parts of the tree most subject to blight has been begun. These are yet incomplete, and can only now be refer to briefly.

The Bartlett and Seckel pears very well represent the extremes, the first being most affected by the disease and the second least. Twigs taken from the tree in February were found to contain 50.2 per cent. and 50.85 per cent. of water respectively. Twigs taken in the same way April 30, bearing flower buds, but with leaves removed, gave 68.7 per cent. and 67.3 per cent. of water. The half-grown fruit, taken the first week in July, gave 79.3 per cent. of water for the Bartlett and 77 per cent. for the Seckel. According to these figures the amount of water in the Bartlett and Seckel twigs during the winter is practically the same, but during growth both the twigs and fruit of the Bartlett contain more water than those of the Seckel. These numbers give some support to the view that succulency and the strength of the disease are directly related, but the data are yet too incomplete warrant a definite statement.

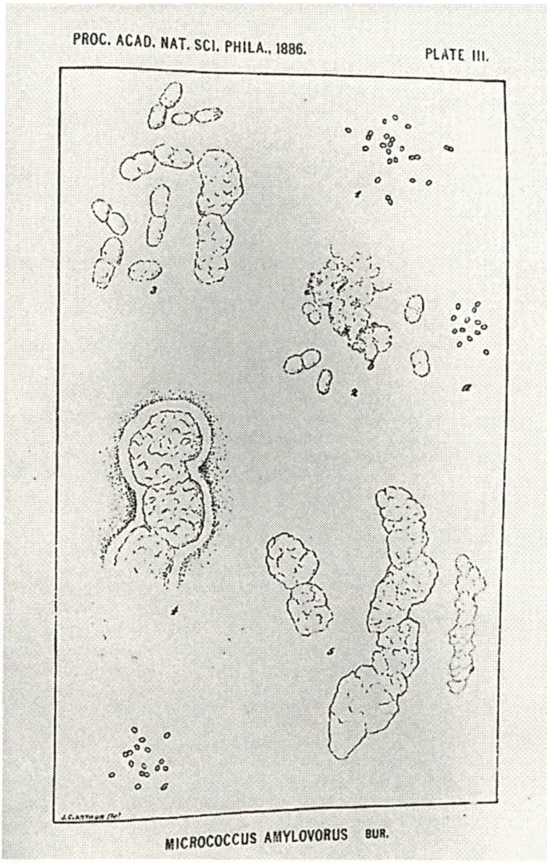

EXPLANATION OF PLATE III.

The drawings were made with a camera lucida and a Spencer's 1/16 objective, homogeneous immersion, of 25° balsam angle. They are uniformly magnified 890 diameters.

Fig. 1 – *Micrococcus amylororus* Bur., grown upon a slice of boiled potato, stained with Bismarck brown and mounted in Canada balsam.
Fig. 2 – From a cultivation in hay infusion: *a*, separate bacteria; *b*, zoogloea. The large mass, only part of which is shown, is made up of smaller masses more or less united.
Fig. 3 – Small zoogloea from a potato infusion, drawn from a preparation in Canada balsam, stained with Bismarck brown.
Fig. 4 – Portion of a zoogloea mass from the same culture, showing an envelope free from bacteria. Drawn from an unstained preparation mounted in glycerine.
Fig. 5 – Three zoogloea from the same culture.
Fig. 6 – From another culture in hay infusion.

Merton Benway Waite

Merton Benway Waite was born on January 23, 1865, in Oregon, Illinois. He was the son of a farmer and early in life acquired an interest in plants. His parents hoped he would become a civil engineer when he entered the University of Illinois, but Waite's fascination with the natural world, combined with the influence of Professor T. J. Burrill, won out. He fell under Burrill's sway and studied plant diseases. After graduation in 1887, Waite continued as Burrill's assistant for one year.[1]

In college, Waite became accomplished in the isolation and culturing of both fungi and bacteria. While he was not present for Burrill's famous work on fire blight, he certainly participated in many of his mentor's later forays into phytobacteriology, including work on diseases of broom-corn and sorghum during which Waite would have worked with poured-plate culture techniques which were relatively new at the time. His ability to work with bacteria was still a rare skill in the late 1880s, particularly among American plant researchers.

In 1888, Waite was recruited by Beverly T. Galloway to come to Washington. Galloway had recently taken charge of the Section of Vegetable Pathology and was looking to expand his scientific staff. Galloway was a remarkable scientific administrator who would oversee the growth of the Section into the Bureau of Plant Industry just after the turn of the century and make it into the recognized center of plant pathology research in the United States.[2] Under Galloway's watchful eye, Waite was the first plant pathologist at the USDA with the assignment to study bacteriology. More to the point, his assignment was fire blight. Blight continued to be a constant threat to pear and apple culture, and there still had been no systematic work on the disease. Arthur had contributed his brilliant papers in 1884-86 and then, like most overworked and overstretched pathologists at public institutions, moved on to other duties. Galloway directed Waite to study fire blight, but he made it clear that his new assistant should devote "special attention to remedies."[3]

The USDA had been founded on the promise of providing practical benefit to America's farmers, but scientists there were somewhat more shielded from impatient constituents than their comrades at state agricultural experiment stations and not at all conflicted by teaching responsibilities like professor/researchers at land-grant colleges.[4] In addition, the Section of Vegetable Pathology was still basking in the glow of the impressive coup it had scored with the introduction of Bordeaux mixture to the United States in the mid-1880s. That "discovery" had demonstrated the value that plant pathology could have to farmers and gave Galloway the budget to hire new people and a grace period to allow his scientists time to work on their research problems. Much of their work necessarily involved testing Bordeaux mixture and other chemical treatments on any and all disease problems, but despite an emphasis on devising spray schedules and improving the mechanics of sprayers, Galloway knew that only an improved understanding of the biology of pathogens and hosts would open the way for future disease control methods. Even so, Galloway expected results from his researchers because he felt pressure from the Secretary of Agriculture who answered to a Congress constantly harangued by their constituents, most of whom were still farmers.[5]

When Waite arrived in Washington in the autumn of 1888, he found that the Section of Vegetable Pathology did not have facilities or equipment for bacteriological laboratory research. His USDA colleagues in the Bureau of Animal Industry had been researching bacterial diseases of animals for some time, including Burrill's old hog cholera friend, H. J. Detmers. Waite borrowed cast-off and extra equipment from the veterinary staff and set up the first phytobacteriological laboratory in Washington. Within a year, additional money allowed the stocking of a "small but excellent bacteriological laboratory."[6] Waite used these facilities to attempt to isolate the fire blight organism using the newest techniques, a feat Galloway considered had "never been accomplished."[7]

Waite's new bacteriological laboratory soon attracted the attention of Erwin Frink Smith. At the time, Smith was working on peach yellows and, as Waite later wrote:[8]

> was already becoming somewhat discouraged by his inability to find any causal organism. . . . (Smith) at once became interested in this bacteriological work and equipment. He had read about these things, but had never had an opportunity to come in contact with them, that is, with such problems as the making of cultures, pouring of plates, isolation of organisms, sterilization and preparation of culture media.

Smith's research into the cause of peach yellows made him into the second phytobacteriologist at the USDA, followed soon by N. B. Pierce who was working on diseases in California that seemed to have bacteria as causal agents. Waite shared the facilities and gave Smith his introduction to phytobacteriology, but Smith would soon outstrip his teacher's knowledge and technique. E. F. Smith would go on to be one of the most important plant pathologists in history and was the man most responsible for setting phytobacteriology on a firm foundation.

In spring of 1889, Waite began traveling to orchards along the east coast to investigate the fire blight situation. He spent the next two field seasons surveying and performing fungicide spray trials as well as engaging the eager cooperation of horticulturists from Georgia to New York.[9] He reported some success with Bordeaux in certain specific cases but could never announce the kind of miracle treatments that researchers working on other plant diseases seemed to be proclaiming every day.[10]

As Waite labored into 1891, Galloway was losing patience with what he perceived as Waite's lack of accomplishment. In early spring of that year, when S. M. Tracy asked Galloway to recommend a researcher who might come to Mississippi to engage in some work on a destructive tomato disease, Galloway was selectively effusive: "Arthur would be a good man but you cannot get him. (Henry L.) Bolley is also a good man and you may be able to obtain his services for a few months. . . . (David) Fairchild is doing splendid work." Then he added, almost as an afterthought, "You can have Waite if you want him. For a job of this kind he would be a good man, but nothing compared to Fairchild."[11] The next month, Galloway wrote to Tracy again and seemed to be eager to have his assistant gone. "Waite will probably not be here after July 1st," Galloway wrote, "so that I see no reason why he can not undertake your work." Waite had only to finish his research on pear blight and Galloway said "it will, of course, be to his own interest to get out as good a report as possible" because "he has never published anything since he has been here."[12]

This last statement is, no doubt, the key to Galloway's dissatisfaction with Waite. Galloway was a consummate bureaucratic politician, and he depended not only on the good work of his researchers, but on the publicizing of that good work to sustain the good will his Section enjoyed with the Secretary of Agriculture. Waite had not yet contributed anything to benefit plant pathology and the Section.

Clearly, that all changed when the results of Waite's fire blight work crossed Galloway's desk. He was never sent to Mississippi and, in fact, by late in 1891 Galloway praised the work Waite presented before the American Association for the Advancement of Science. After two years of effort, Waite unveiled proof of insect transmission of fire blight, the first scientific evidence of insect transmission of any plant disease.

Waite had several years experimenting on pear trees on the grounds of the USDA in Washington as well as in cooperators' orchards in Maryland and New York. In early spring, he inoculated these trees with fire blight bacteria to create what he referred to as an "artificial epidemic." His studies led him to conclude that the bacterium multiplied in the blossoms, specifically in the nectaries, and traveled down the pedicels into the stems. In fact, Waite's experiments did not demonstrate the exact method of blossom infection, but that was not his primary goal.

Waite then turned to the issue of how the disease moved among trees or even among different blossoms on the same tree. He artificially infected certain blossoms on pear trees and used bags or nets to protect other blossoms from contact with insects that visited the infected flowers. He observed that the bagged blossoms did not blight while others in the same orchard or on the same tree where insects did visit showed signs of the disease.

From the blighted blossoms, he collected bees, excised their mouth parts, and extracted bacteria. When these bacteria were cultured and studied, they appeared to be *Bacillus amylovorus.* When it was then used to inoculate pear trees, fire blight resulted. Therefore, Waite proved that insects, in this case honey bees, transmitted fire blight. He later showed that other insects played similar roles, but bees remained the primary vector because of their pollinating duties. Waite also specifically, and incorrectly, ruled out the possibility that the bacterium could be spread by the wind.[13]

The peculiar fact was that Waite still did not publish the results of his research in a scientific paper that detailed the premise of his experiments, their progress, and his conclusions. His epochal work was presented in a single paragraph abstract in the AAAS *Transactions* and *The Botanical Gazette.* It was summarized in a general article in the *Transactions of the Peninsula Horticultural Society* where Waite spoke in early 1892, and it was hyped briefly by Galloway in the USDA *Annual Report.*[14] But Waite apparently still resisted writing up his results as an official scientific paper.

This disinclination by Waite to trumpet his discoveries was surprising given the clear significance of this work to the larger world of biology. It was the first firm evidence that insects played a significant role in the transmission of plant diseases. Waite's research coincided with the discovery of insects as vectors of human/animal diseases. In 1893, Theobald Smith and Frederick Kilborne of the USDA Bureau of Animal Industry published on the role of ticks in the transmission of Texas cattle fever. Waite preceded them to publication, such as it was, but some of the original work on Texas fever by Dr. Cooper Curtice started around the same time Waite began his studies of fire blight.[15]

Waite and the scientists at Animal Industry did not originate the concept that insects carry disease in both plants and animals. Arthur, among others, had discussed the possibility of insect transmission of blossom blight in the mid-1880s. And in terms of Texas fever, ranchers often associated tick infestation with the disease, but scientists largely dismissed those observations. However, Waite and his veterinary colleagues were the first to demonstrate by scientific experiment the fact of insect transmission of disease. There is no reason to believe that Waite and the veterinary scientists coordinated their work in any way. Waite first became interested in proving the possibility of insect transmission of fire blight in his first year at the USDA while observing the actions of insects in an orchard in Georgia while at the same time Cooper Curtice was beginning to study the role of ticks in Texas fever.[16] It is not unlikely, however, that they were aware of each other's research given the relatively small size of the USDA at the time and the fact that Waite had contact with the Animal Industry bacteriologists when he had begged equipment from them to establish his plant bacteriology laboratory.

It is likely that Waite did not parade his insect transmission work of 1891 as a discovery of great impact because he did not consider the work completed. He had yet to demonstrate what implications, if any, insect transmission had for disease control. Since it had been determined that insects carried fire blight from tree to tree, "this led to investigations to find out whether the insects could be dispensed with."[17]

The answer was no. After several more years of research, Waite found that bees were required in orchards to ensure proper fruiting of most varieties of pear. Even more importantly, during the course of the work he found that pear trees, by and large, did not self-pollinate. This meant that orchards of mixed varieties as well as the use of honeybees to facilitate pollination were advisable. This was a major systematic breakthrough for pear growers and helped define modern horticultural practice in terms of pear orchards.[18] Although Waite was justifiably proud of these contributions to horticulture, the pathological implications of the insect work were disappointing to him given the fact that chemical treatments had only a limited success for the control of fire blight.[19]

Waite continued to study the biology of fire blight and search for better disease control techniques. While he focused on blossom blight, he also uncovered one of the most important parts in the disease cycle, namely the source of inoculum for infection in the spring. He had dismissed the possibility that the bacteria could survive for long in the soil, a belief that Arthur had propounded. Waite observed that the pathogen overwintered in the tree in cankers and this "hold over blight" could provide a ready source of initial inoculum in the spring.[20] He forcefully advocated orchardists to search for signs of "hold over blight" and cut them out wherever possible during periods of dormancy "for it affords opportunity to strike the enemy at a disadvantage."[21]

Waite's laboratory work also resulted in the first true description of the *E. amylovora* in culture. In addition to stating flatly that the bacterium "does not decompose starch," he described it as an "oval rod like bacillus. . . . Has no capsule, but is supplied with several flagella scattered over the surface."[22] This was the first observation of flagella on *E. amylovora*, although Arthur had noted that the bacterium was motile in culture without seeing flagella. Smith later claimed that the flagella were difficult to stain. Smith also credited Waite with being the first scientist to give a detailed, thorough, and correct description of the fire blight pathogen.[23]

In the course of his careful culture work, Waite drew a distinction between the fire blight bacterium and a similar bacterium, the so-called "false yellow germ," often found in conjunction with it. This bacterium was later named *Erwinia herbicola*. Waite claimed that the fire blight bacterium was always white, and any sign of yellow was a sign of "an impure culture." That provides evidence that Arthur may well have cultured *E. herbicola* in some of his earlier work. At the very least, some of his *E. amylovora* cultures were likely conta minated with *E. herbicola* because he noted cultures that were yellow. Arthur later admitted that he had "confounded in my published account a non-pathogenic with the true pear blight bacillus" and agreed that Waite's published description of *E. amylovora* was the standard.[24]

Waite's expertise on fire blight was called on during severe outbreaks of the disease in Colorado and California in the early 20th century. The California epidemic was particularly troubling. That state had already become a major pear producer and had never yet suffered from fire blight. It was thought the climate was not conducive to the disease. However, just after the turn of the century, fire blight hit California hard. Some estimates claim that two-thirds of the state's Bartlett pear trees, which was the most widely grown and economically lucrative variety, were destroyed.[25] Waite eagerly worked with California officials on an ambitious eradication program. The problem, however, proved too large because the state's vast orchards required a level of inspection for which there was no funding, and Waite encountered a certain, natural level of suspicion and hesitance on the part of orchardists to change their cultural practices or their varieties.[26]

Despite his years of research and work on fire blight, Waite never acquired the level of scientific legend such as that shared by Burrill and Arthur. He remained the bureaucrat scientist laboring away in the service of the USDA. Perhaps it is because the demonstration of insect transmission is not granted the same status as a "discovery" as the identification of the causal agent. It is likely also that the publication of the insect transmission work was buried in papers dedicated more to pollination studies or to practical papers on the control of a disease that had no controls. Interestingly enough, when Erwin F. Smith wrote his magisterial two-volume work, *Bacteria in Relation to Plants* (1911), he seemed concerned to establish Waite's priority over insect transmission. After describing Waite's method of inoculating pear trees, capturing bees from infected blossoms, removing bacteria from the bees and culturing it, and then reinoculating trees, Smith added a personal note:[27]

> Everybody connected with the plant pathological work of the U.S. Department of Agriculture at that time had knowledge of these results. The writer, among others, saw all of the experiments described and knows that they were well done and that the above brief outline can be accepted as an accurate statement of what actually took place.

Waite himself was never overly concerned with establishing his name and carving out a place in the science. He was more interested in doing the work at his own pace. He was universally well liked by colleagues and regarded as someone who willingly shared whatever time and expertise he had with others. Indeed, he played a formative role in making E. F. Smith into the premier plant bacteriologist in the United States if not the world, constituting a semi-direct line of bacteriology mentorship from Burrill to Smith. Smith, therefore, took care to protect Waite's reputation even if Waite would not do so himself. And more, it was in Smith's influential writings in the early 20th century where Burrill's place as the father of phytobacteriology was cemented to the exclusion of Arthur or any other scientist.[28]

Waite continued to serve as the USDA's point man on fruit tree diseases throughout his career, which stretched for more than three steady but relatively unremarkable decades, after his important discoveries of the 1890s. As with many scientists, as his career advanced, he moved increasingly into administration and away from research. Waite's knowledge of chemical disease control methods made him a useful member of the Federal Insecticide and Fungicide Board, and he later served on the Federal Horticulture Board which administered the early quarantine regulations. In 1919, he was awarded an honorary Doctor of Agriculture by the University of Maryland, his only graduate degree. M. B. Waite died in 1945.[29]

It is an unfortunate result of Waite's failure to publish fully at the time of his discovery of insect transmission that only a brief abstract of his AAAS presentation, titled with typical modesty "Results from Recent Investigations in Pear Blight," is left to us to represent the initial presentation of this groundbreaking work. That abstract, although unsatisfying, must serve as the earliest publication of his important research on insect transmission. Waite continued for more than a decade to publish his fire blight research piecemeal, and this is reflected in the mismatched collection of papers reprinted here to represent his important contributions to the foundation of fire blight studies.

The second piece is an excellent example of Waite's skill as a laboratory scientist. The short paper for AAAS, "The Life-History and Characteristics of the Pear-Blight Germ," has the best and most definitive description of the bacterium up to that point. Waite succeeds in clearing up a number of misconceptions and falsehoods that had spread through the literature.

Waite is no doubt best known for the often-cited "The Cause and Prevention of Pear Blight," published in the USDA Yearbook for 1895. Although this is a fine piece as far as it goes, it says little about his work, opting for a rather obvious retelling of the facts. "The Cause and Prevention of Pear Blight" does not give a sense of Waite's work, it does not outline his experiments with insect transmission, and, given its likely familiarity to those knowledgeable about the field, we have omitted it here. A useful discussion of his work in the early 1890s came only a decade later with "Relation of Bees to the Orchard" in 1902, and therefore we have chosen this little seen piece for reprinting here.

Finally, we have chosen another obscure Waite article from 1904, "Pear Blight and Its Treatment – Life History of the Disease," published by the State Board of Horticulture of the state of Colorado, based on a presentation given by Waite during one of his many western tours after the disease struck with such vigor in that region. This paper, little seen or read since its publication, is a good example of a full-scale article by Waite, as opposed to the smaller pieces already presented. It is an excellent overview of the disease from both a scientific and agricultural point of view. It gives a good sense of Waite's versatility as a scientist and his ability to communicate his knowledge to specialists and practitioner alike. It is an excellent example of plant pathology at the dawn of the 20th century.

This paper also provides a fitting conclusion to the formative years of fire blight studies by showing how the disease, although better understood in the quarter century since Burrill's pioneering studies, is no closer to being under control and, in fact, is spreading with great rapidity and causing as much alarm to fruit growers of the early 20th century as it did to their predecessors of the 19th or 18th centuries.

LITERATURE CITED

1. Roberts 1946, 175.
2. Peterson and Campbell 1997, 29-43.
3. Galloway, B. T. to W. J. Beal. August 2, 1889. USDA Letters Sent, 1885–1888.
4. Rosenberg 1976.
5. Campbell, Peterson, and Griffith 1999, 159-180.
6. Galloway to Beal. August 2, 1889. USDA Letters Sent, 1885-1888.
7. Galloway, B. T. to F. S. Earle. July 22, 1889. USDA Letters Sent, 1885-1888.
8. Waite, M. B. (n. d.) When Dr. Erwin F. Smith became interested in bacteriology. In Smith Papers. 1865–1940.
9. Waite, M. B. to O. P. Bennett. June 28, 1889; Waite, M.B. to C. G. Bestle. July 5, 1889. USDA Letters Sent, 1885-1888.
10. Waite 1892b, 33-34.
11. Galloway, B. T. to S. M. Tracy. March 14, 1891. USDA Letters Received 1891-1900.
12. Galloway, B. T. to S. M. Tracy. April 7, 1891. USDA Letters Received 1891-1900.
13. Waite 1892a, 315; Waite 1892b, 33; Waite 1902, 66-70; Smith 1911, 55.
14. Waite 1891, 259; Galloway 1892, 373; Waite 1892a; Waite 1892b.
15. Logue 1995, 1-53.
16. Rodgers 1952, 210fn.
17. Waite 1895, 96-99.
18. McGregor 1976, 83.
19. Waite 1894.
20. Waite 1898b, 785.
21. Waite 1896, 298.
22. Waite 1898a, 427-428.
23. Smith 1920, 373.
24. Rodgers 1952, 116-117; van der Zwet and Keil 1979, 6, 62.
25. Bonn and van der Zwet 2000, 39.
26. Smith 1911, 191-192.

27. Smith 1911, 55.
28. Baker 1971, 622.
29. Woods 1945, 632; Roberts 1946, 177.

Fig. 5. M. B. Waite making a transfer of the fire blight bacteria in his office at the USDA. (Courtesy of the National Agricultural Library, Beltsville, MD.)

Fig. 6. M. B. Waite conducting fire blight inoculations on pear in July, 1893, on the grounds of the USDA, Washington, DC. (Courtesy of the National Agricultural Library, Beltsville, MD.)

Results from Recent Investigations in Pear Blight.*

M. B. Waite

Pear blight is a disease which works only in meristematic tissues. Twig blight and blossom blight explained. Methods of isolation of the microbe of blight and its cultivation on nutrient culture media. Some characteristics of the germ of blight. It is a motile *Bacillus*. Inoculation experiments with pure cultures of the Bacillus of blight. Blight in the flowers and experiments with the blight germs on pear flowers. The blight bacteria grow in the nectar and multiply there as saprophytes and then enter the tissues. The Bacillus of blight in the nectar is carried from flower to flower by insects visiting the flowers for pollen and honey. An artificial epidemic of pear blight started by infecting a few trees on the edge of an orchard and allowing free access of insects. Mosquito net bags over the flowers keep out insects, therefore protect the flowers from blight. Experiment with bees and blight germs under a tent. Trees infected with blight germs artificially may be sprayed with fungicides and the greater part of the damage prevented.

*Originally published: 1892. *Proceedings of the American Association for the Advancement of Science*. Salem, MA: AAAS. 315.

The Life-History and Characteristics of the Pear-Blight Germ.*

M. B. Waite

Beginning in the spring the germs on the new growth of the season first appear on the nectar disks of the blossoms. The bacilli live and multiply in the nectar and are able to enter the nectar glands without a puncture or injury and thus normally get inside their host. The distribution from flower to flower and tree to tree is through the agency of insects, mainly flower-visiting sorts. Infection also occurs on the young shoots and less frequently on the fleshy bark through injuries. Insects and birds are agents of distribution and inoculation in these cases. No evidence could be found that the germs are carried by the wind. The blight germs usually die out in the twigs which are blighted and dead but in certain cases the germs manage to keep alive through the summer by making slow progress in the fleshy bark. Such cases may succeed in living over winter. Winter weather is favorable to the longevity on account of the moisture and low temperature. The cases of "hold over" blight start off again in spring and exude quantities of gummy matter full of the bacilli. This is visited by insects, especially flies and wasps, and carried on to the newly-opened flowers thus completing the life cycle.

The condensed characters may be given as follows: An oval rod like bacillus .6–.8 μ by 1 to 6 μ long. Constant in diameter but varying greatly in length. Occurs singly or in young cultures in pairs, chains or masses. Stains readily with the ordinary aniline dyes either watery or alcoholic solutions. Has no capsule, but is supplied with several flagella scattered over the surface. It is actively motile. Does not produce spores. On nutrient beef and potato broth produces first a strong turbidity and a slight granular pellicle on the surface, which breaks up and settles to the bottom. The color of the mass is milky white on all solid media.

On agar plates the outside colonies at ordinary temperatures 18–20° C. reach a diameter of about 1 mm. In forty-eight hours and at the end of a week become 5 to 6 mm. across. A temperature of 36–37° C. starts the growth more promptly but results in a feebler ultimate development.

*Originally published: 1898. *Proceedings of the American Association for the Advancement of Science.* Salem, MA: AAAS. 427-428.

The addition of malic or citric acids in small amounts so as to acidify the agar increases the vigor of growth while an excess of alkali diminishes it. On gelatine made from the commercial brands, the opposite effect is produced. Gelatine should be neutral to phænolphthalein to insure vigorous development. There is a moderate liquefaction in good gelatine cultures. A moderate growth is made on sterile potato cylinders.

In the fermentation tube it decomposes sugar without the formation of gas. It is most vigorous on maltose the cultures becoming strongly acid, and is slightly less so on cane sugar, dextrose and levulose. It is aërobic and facultative anaërobic. It produces no pigment or coloring matter of any sort and no odor. It does not decompose starch. Its principle food consists of nitrogenous matter, sugars and probably, to some extent, certain organic acids, the very substances which occur in vigorous, young, growing tissues of the host. Certain statements formerly made are now known to be erroneous.

The germ mass is said to be yellowish white on potato. This could only come from an impure culture as the true pear-blight germ is always white. Gas, or in some places CO_2 gas, is said to be formed. This never occurs. Butyric acid is said to be one of the products of its decomposition. The germ produces acids but never butyric. Starch is said to be decomposed and used as a food, but so far we have never been able to demonstrate this. The germ is said to live over winter in the soil. In our search we have failed to find it in such places and its life cycle is complete without it.

Relation of Bees to the Orchard.*

M. B. Waite

I will discuss the subject under three heads:

First, bees and the monilia fungus.

Second, bees and pear blight.

Third, bees as pollenizers.

Bees are certainly intimately connected with the orchard, both in health and disease. Taking up the first subject, I think I can give you something new. It is probably the least important of the three matters which I shall take up, but it is the newest. I am quite well satisfied from observations made during the early part of the present summer that bees are largely responsible for the rapid distribution of the peach and plum rot fungus. They are not the only culprits connected with that distribution. Wasps and soldier bugs and several other puncturing insects are also concerned in the matter, and the wasps and puncturing insects are the ones which usually make the openings, especially in perfectly sound fruit. However, bees rapidly follow in the punctures by other insects. In June and July hundreds and thousands of bees may be seen swarming over the early ripening peaches and Japanese plums, and in going over the rotten fruit they carry the spores into the wounds made by the wasps and soldier bugs, and into weather cracks. The peaches cracked open badly by the wet weather this year.

Oldmixon was the worst in Maryland, but even varieties like Elberta showed a great many weather cracks. Into these cracks the bees penetrated. The even ate large holes in the fruit. I have seen a hole large enough to hold three bees inside of it, and even on gathering the fruit, they were so busily engaged that they were still at work inside the hole.

I have endeavored to follow up this subject a little more closely, and tried to find out how much the monilia fungus could spread without the aid of bees, and investigation showed that it was able to spread without the aid

*Originally published: 1902. *Proceedings of the Twenty-seventh Session of the American Pomological Society.* Lansing, MI: American Pomological Society. 66-70.

of insects; in other words, it does blow by the wind, and gets about in that way. I covered trees with mosquito netting, and also branches, and thought perhaps I might be able to prevent the spread of monilia within the mosquito net protection, but was not able to do so entirely. I somewhat reduced it, however, showing that the bees are active agents in the distribution of monilia.

Taking up the second subject of bees and pear blight, we may say that bees are very active agents in the distribution of pear blight. In order to fully understand, however, the exact part that bees do play in the distribution of pear blight, it will be necessary to go very briefly over some other points concerning this disease, and to give, in fact, the life history of the pear blight bacillus through the year. In the first place, we have three methods by which the pear blight germ enters the tree: First, through the blossoms, resulting in what is commonly called blossom blight, to distinguish it from the other methods of communication of the disease; Second, through the tender tips of growing twigs, but either blossom blight or twig blight may run down onto the limbs and trunk; and, Third, blight may enter directly into the fleshy bark of young trees.

The latter is comparatively a rare method, in fact by far the greater number of infections are directly through the blossoms. Now, of course, it is only in blossom blight that the honey bee is concerned, and furthermore, all insects which visit the pear and apple blossoms and quince blossoms are more or less concerned in the matter of distribution on the blossoms. I may say also that other insects than the honey bee are largely instrumental in the distribution of pear blight on the twigs, and we have scarcely been able to find a direct, normal method of introduction of the disease in the twigs without the intervention of some mechanical or insect puncture. Beginning with the spring of the year, the disease breaks out on the blossoms. Ordinarily, trees do not have the blight very badly until they come into bloom and this, as I said before, is mostly due to its distribution by bees and other Insects which visit the flowers and are the main factors. The pear blossom is a very open one and is very extensively visited by a whole list of insects. I started out to get a list of insects which visit pear blossoms, but when I reached the number of forty, I gave it up. Nearly all the flying insects, the bee being most active of all, but even beetles and wasps, and occasionally even a humming-bird (the latter of course not being an insect) visited the pear blossoms and carried the disease along.

The blight begins on the blossoms in early spring; then during the summer we may have twig blight clear into September. The disease runs down on the twigs, and in the great majority of cases dies out. The disease works in the bark and as a rule finds the tree, at some time during the late spring and summer, a little too dry and tough for it to take hold of, and it dies out. Occasionally infections, however, keep running down on the twigs, get into the fleshy bark, and keep on working slowly till fall. After this season of the year, the trees are so moist that these germs will not dry out but will live over winter, resulting in what I call "hold-over blight." It stands zero weather perfectly. In the spring, those hold-over cases start off when root pressure begins and when the tree is forced full of sap. In some instances in the South, very active changes go on in these hold-over cases during mild winter weather, but in the Northern States they don't have the time to do much damage. The virus runs out on the sides of the tree and drops down. I saw some very fine cases in the mountains of Virginia last spring. The flies and wasps and other insects swarm on this exuding sap, and as the blossoms open they fly from that to the opening blossoms and start the infection for the next year.

We have in succession, first, blossom blight, then blight running on the branches, then hold-over blight, then re-infection the next spring, through the blossoms. When it is started on the blossoms, it is carried like fire. Now, as I started to say a moment ago, you would like to have the authentic proof that bees carry pear blight. I worked long and patiently on this problem for three or four years. In the first place, the great abundance of blossom blight leads to a suspicion, at least, that we have here some normal, effective method of distribution. There is some regular way about it, and we shall see what that method is. In the second place, the disease begins in the nectaries. The germs of pear blight are found growing and multiplying in this nectar. I determined this mostly by artificial infection. By taking pure cultures of the pear blight germ and touching the nectary with a camel's hair brush I started the pear blight off. It requires no puncture. It is the only point on the tree where the germs enter normally without a puncture. The nectary has no shield or cuticle over it to keep out the germs. Now, my first acquaintance with this relationship of bees to pear blight came about something like this: When I had succeeded in isolating the pear blight germ, I immediately tested the validity of my cultures by a series of experiments with a camel's hair brush on some blossoms, and I inoculated and labeled a number of blossoms around the lower part of a pear tree on the Agricultural Department grounds at Washington. I spent a good deal of time actually watching those blossoms growing, and of course here came the bees right in front of me, and as the germs began to grow in the nectaries, they began to visit my infected blossoms, and I saw them fly from these over the trees to the other blossoms. I captured two bees, caught them in the act of visiting the infected blossoms, and taking them into the laboratory made plate cultures by the method in which we ordinarily isolate bacteria, and succeeded beautifully in isolating pear blight germs from the mouth parts of the bees.

In other words, we actually caught the bees in the act, and this was repeated a number of times and in different parts of the country. Now, it seems as though we had the matter pretty well proven. My trees that I had infected came down beautifully with pear blight. The clusters all came down which I had infected and labeled, and a large number of others all around the tree.

The question then was as to what other methods of distribution the blight germ might have, and I set about with quite an elaborate series of experiments, to find out the truth of the matter. The main question was whether the pear blight germ was carried by the wind. If it could be blown by the wind, it would not make any difference whether the bees carried it or not, but if it was not blown by the wind, it became a matter of considerable importance. So, having gone through one spring in Washington, I moved up to Western New York and passed through another spring the same year. A gentleman living near Brockport, N. Y. very kindly turned over a small orchard to me in which I could start a small, artificial epidemic of pear blight. We inoculated about six or eight trees, and about eight or ten small branches on each tree. Before inoculating them, I covered a large portion of the trees with bags of various density, mosquito netting, cheese-cloth and paper, reasoning that by covering the trees with mosquito netting, it the disease was blown by the wind, it would certainly blow through the mosquito netting and would appear under the bags.

The disease gradually spread from my centers of infection, but in no case did the disease get through the mosquito netting, although a large percentage of the clusters broke down, on the trees that were infested, outside of the mosquito netting. We inoculated blight in the bags and kept it in; we inoculated it outside, and kept it out. Now, in further support of the view that I have taken that pear blight is not borne by the wind, we have two important facts: The first is that the pear blight virus is always a sticky mass, which cannot be blown by the wind. It is a gummy material that can readily be carried by anything that touches it, but not by the wind. The second fact is that the pear blight germs on drying rapidly die. In ordinary orchard conditions, a few days after the twig dries out the germs die. So that pear blight germs are not lying around promiscuously. We have no evidence whatever that we can get infection in any other way. So I think you will be willing to accept my proposition, that bees are active agents in carrying pear blight on blossoms, and that the blight is not carried by the wind.

Now, we have been giving the honey bee such a bad record that it is with great pleasure that I turn to another phase of this subject. Very early in this work, the question came up in my mind, if the honey bee was doing this unfortunate work, what was its function on the blossoms? The question came up, were our pear blossoms of such a nature that they required this insect fertilization by the bees? So I started out to determine the relation ship between the bees and the fertilization of the blossoms. The result was a long series of experiments lasting over three years.

On account of the limited time, I shall have to state very briefly, the results of this pollenization work. In the first place it came out that most of our pears, as the result of careful hand pollination on tests, are sterile to their own pollen. Cross-pollination is characteristic of our orchard fruits. It is perhaps not true of peaches and quinces, but as far as apples and pears are concerned, cross-pollination is a necessity.

By trying a large number of experiments, I concluded that insects were the largest agents in cross-pollination. The blossom of the pear is not of the type of wind-fertilized blossoms, but is distinctly of the insect-fertilized type. Now, this matter is so complicated that it will be impossible, in the short time at my disposal, to go into the details of fertilization. To make the story short, we may say that from the biological standpoint, the bees are doing their normal, natural work in visiting the pear and apple blossoms. The blossoms are for the bees to pollinate, are developed by insects, and the insects have been developed in correlation to them. It is a normal, biological process, this visiting of orchard fruit-blossoms by bees. The bees are there performing their proper function.

Now the question is, is the honey bee to be looked upon as useful or as injurious? Do its beneficial effects equal its injurious effects? When there is no pear blight about it, it is easily decided in favor of the usefulness of the bee, but when there is an outbreak of blight, it is a rather hard question to decide.

There are no doubt times when pear blight is rampant in an orchard, when it would be better not to have the honey bee or any other insect in there until the blight is past. On the other hand, from the standpoint of the beekeeper, it can be said that in most cases wild bees and other insects would do a large part of the work, both good and bad, without the honey bee, so that we could not get rid of the infection of the blight even if we did dispense with the presence of the bee. I feel satisfied that in the Eastern United States the honey bee must be looked upon as a useful element in the orchard. I have an apiary in my own orchard. The people in California have had some terrific outbreaks of pear blight and are very much concerned over the matter, and they have reached the conclusion that they can dispense with the honey bee as an orchard pollenizer; that their Bartlett pears will set all the fruit that is necessary, without insect fertilization. I have never investigated the subject in California. My only knowledge of the case is due to my observation of California fruits in the markets. There are many seedless Bartlett pears from California that it looks as if they were necessarily self-pollinized, and there may possibly be some truth in their view that their pears will set without cross-pollination. In that case, it moves the thing a little bit against the bee, and if pear blight is rampant in an orchard, it would certainly be desirable, – at times at least – not to have the bee about.

In conclusion, I may say that I have had to just barely touch upon matters of extreme importance that should require an evening for their discussion, but I am prepared to defend my propositions by further explanation, if you find it necessary.

Pear Blight and its Treatment – Life History of the Disease.*

M. B. Waite

Pear blight appears in spring on the blossoms. Occasionally it attacks the young shoots before the flowers open, but ordinarily the first appearance of the disease is in the nectaries of the flowers. The closest examination is necessary in order to detect the first appearance of blight on the blooms, but the disease develops in the nectar, enters the tender exposed surface of the nectaries, extends downward into the tiny fruit which forms the base of the flower, and from this enters the flower stem, eventually spreading into other parts of the flower cluster, killing the whole cluster with its attached leaves. It may extend from the cluster into the fruit spur and so on into the twigs and branches. The blight germs are carried extensively from flower to flower, from tree to tree, and from orchard to orchard by insects, which visit the flowers in search of honey and pollen. The most active insect and the one which does the most distributing of the blight is the ordinary honey bee. Wild bees, wasps, flies, many species of beetles, and other insects, and in some cases even humming birds, visit the blossoms of apples and pears and thus carry the blight germs about.

The effects of blossom blight are at first not conspicuous, during the period of from one to two months following blooming, after the disease has spread into the fruit spurs and branches and the affected leaves and twigs have dried out and blackened, the widespread damage becomes plainly apparent. The period of blossom attack is necessarily very short, but during this short interval on an average more than 95 per cent. infection by pear blight takes place. In many orchards this is almost the only method in which the blight enters the tree.

However, there are at least two other modes of entry of the pear-blight germ. Under favorable conditions the disease is able to enter the tender tips of growing twigs. Here again insects play an important part and probably do almost all of the work of inoculating the germs into the tree. Except under

*Originally published: 1904. *Annual Report of the State Board of Horticulture of the State of Colorado.* 38-49.

very unusual conditions, pear blight is unable to enter the twigs without a puncture or wound of some sort. The breaking off of a tender young leaf by the wind, or the scratching of one twig against another may make a wound in which the blight can enter if the germs are present. On the other hand, pear blight is not blown about by the wind. The virus is a sticky, gummy exudate from the tree, and ordinarily it requires transportation by some living thing, such as a bee or other insect or a bird or some other animal which can move about.

A third method of entry of the pear-blight germ is directly into the fleshy bark of the larger branches or body of the tree. This green fleshy bark in the proper season is susceptible to the disease, but ordinarily it is so protected by the cuticle or the thin layer of cork on older branches that the pear-blight germ can not enter without a wound or puncture. In several cases under observation by the writer, sap-suckers in puncturing the trees have been found to distribute the disease. We may safely assume, I think, that where blight appears on the body or the branches, the germs have been inoculated into the tree by some outside agent. The nectaries, therefore, are the only normal surfaces in which the pear blight regularly gains entrance into the tree without a puncture. I am assuming that the bacterial nature of this disease is well understood. The disease once entering into the tree attacks the tender growing tissues of the blossom clusters and young twigs as above mentioned. It then, under favorable conditions, spreads downward into the fleshy bark of the tree, usually killing not only the outer bark in where it thrives best, the soft bast or inner bark, which represents the active living part of the branches and body of the tree. The killing of this layer between the bark and the wood, therefore, means the destruction of the branch or body at that point. Occasionally the germs grow in the vessels of the wood of the twigs, branches or body of the tree, but this is somewhat exceptional. If the cambium is killed on a branch, the branch is, of course, doomed.

It will not be possible to go into all the details of the manner in which the blight spreads in the tissues, and the numerous and sometimes curious ways in which it girdles the branches or even the bodies of quite large trees, but in all these cases the germs of pear blight are responsible for the trouble and are developing according to the conditions of the tree and the state of their environment. In a majority of cases the germs find themselves under conditions, after killing a few blossom clusters or twigs, that are not suitable for their further development; the activity of the blight ceases, the twig dries up and a layer or crack forms, separating the dead wood from the live. The disease is purely local, killing completely every cell as far as it extends in the tree, but leaving all below this uninjured. The discoloration at first when the disease enters the tissues is rather slight, the first visible effect being a water-soaked or soggy appearance of the tissue. This changes in a day or so to a brownish tint and in five or six days, at least in the summer time, to a dark-brown discoloration. When the crack or definite line of demarkation (sic) between the dead bark and the live is plainly evident the disease ordinarily has stopped. When the line of demarkation (sic) has not formed and the blight blends off into the healthy tissues, the disease is still active and progressing. In the great majority of cases, fortunately for the orchardist, the disease comes to a standstill, the twigs dry up, and the pear-blight germs all die. The organism is a very delicate germ, rarely living in summer time more than two or three weeks in any particular twig or branch after it has killed that particular portion of the tree. In order to survive, the germs must keep pushing onward. In the fall the year certain cases of the disease have been slowly developing and still remain active. After the trees shed their leaves and go into winter condition, the cases of active blight which have survived up to that period are almost certain to live over the winter in the tree. Cold does not kill the pear-blight germ, but retards its activity. The mild weather of early spring starts the disease into renewed growth. The tree being gorged with sap forced upward by root pressure, the germs develop with renewed activity. Gummy matter, consisting of a mass of bacteria mixed with the broken-down tissues and the sap of the tree, is exuded from the blighting branches. These "hold over" cases, therefore, furnish the starting point for re-infection of the blossoms in the spring. Flies and other insects visit gummy matter and carry the germs on to the opening blossoms.

CONDITIONS GOVERNING THE SEVERITY OF THE DISEASE

We may best get our bearings in controlling this disease by a study of the conditions which govern it. Pear blight behaves in such a great variety of ways and is influenced by so many different factors that it is not always possible to determine the reason for the outbreak or for the particular severity of its attack at a certain time and place. Sometimes in the East the disease dies out over whole sections or counties. Then again it may reappear as virulent as ever after an interval of several years. We may best discuss these conditions from two standpoints: A – those relating to infection by the germs; B – those relating to the development of the disease after infection. As you doubtless know, we may have numerous infections, perhaps 500 or 1,000 cases, scattered over an apple tree or a pear tree, with nothing more than the fruit spurs or a few twigs killed as the result. On the other hand, a single infection of a young pear tree may frequently develop into a severe case which runs clear down on to the body and kills the tree.

A. – Factors Relating to Infection by the Germs. – In the first place, it is necessary for infection of pear blight to have the germ causing the disease present in the vicinity. No matter how favorable the conditions may be, unless the pear-blight germ is present in the orchards no disease will result. If only a few germs are brought in from some outside source during or toward the close of the blossoming period, a few cases only may result the first year. These perhaps holding over in the orchard may give the disease such a start that during the second and third year it spreads with great virulence over the trees.

The second factor largely influencing infection is the weather conditions, especially atmospheric humidity. In the East, muggy or cloudy weather has been repeatedly observed to be extremely favorable for the development of new cases of blight. Many people supposed, not without reason, that the irrigated orchards of the Western states, on account of the aridity of the atmosphere, furnished conditions so wholly unsuitable for infection by pear blight that this factor alone accounted for the absence of the disease in early days. On the other hand, we now feel certain that it was the first factor, namely, the absence of the germ in the neighborhood, which decided the matter. I can not understand fully how it is that pear blight is able to infect the orchards as thoroughly as it has done in Colorado. Of course the blossoms with their nectaries are so susceptible that no amount of dry weather can make them immune to the disease, but we must accept the facts as we find them, namely, that during the spring and summer there are conditions – probably foggy and cloudy infection periods, which you may be able to observe – in which the disease can get its start without drying out.

The third factor is the presence of insects. This is a very important question and is ordinarily one over which the orchardist has very little control. The effect of the honey bee in distributing blight has brought up a very serious question as to the advisability of having bees in the orchard or within three or four miles of important commercial orchard sections. The honey bee is certainly the most active agent in distributing the blight, and where it is possible, it certainly the safest plan to temporarily remove the bees from the immediate vicinity of orchards while the trees are in blossom. In the Eastern states there are so many wild bees and other insects that probably infection would take place largely even though all the bees were removed or exterminated. Furthermore in the Eastern states the honey bee is an important factor in the pollination and the setting of the fruit. Without him ordinarily we can not expect a full set of apples and pears, as the writer has demonstrated by an exhaustive series of experiments. In Colorado I have been assured by many orchardists, and my own observation is in corroboration, that the trees set too much fruit; that the conditions are so favorable that self-sterility is a matter of small importance. At any rate, this factor of insects as pollinizers (sic) of fruit should be taken into account in considering this problem.

The fourth factor governing infection is the presence or absence of blossoms. Trees ordinarily are not infected by the disease unless they are in bloom. Nursery trees or young orchards may be occasionally attacked, but the blossoms, on account of their attraction of insects carrying the disease and on account of the facility with which the germ enters the nectaries, become the principal source of infection.

B. – Factors Relating to the Spread of the Disease After Infection. – (1) The Species and Variety of the Fruit Tree. – Practically all pomaceous fruits, not only the cultivated but the wild species are attacked by this disease. We have no cultivated pomaceous fruits that are truly free; in other words, there are no blight-proof pears. However, the resistance which the host affords to the parasite varies enormously. There are, therefore, all degrees of relative immunity from the disease. As a rule, pears are attacked more seriously than apples or quinces, although certain kinds of the latter may be killed outright under circumstances favorable for the disease. Some varieties of pears blight so hadly (sic) that they can not be cultivated at all. The Idaho is a notable example of this sort. Neither can Clapp's Favorite be grown in the South Atlantic states. The Bartlett, which is the greatest commercial pear of Colorado, is, unfortunately, quite susceptible to the disease. While it is not the worst tree to blight, it belongs among the group of badly-blighting varieties. As a rule, slow-growing varieties like the Seckel and Winter Nelis are less seriously attacked, although a standard old pear, the Duchess, which grows quite vigorously and produces large fruits, is one of the relatively immune varieties. Among the apples, the Russian varieties have proved the worst blighters. Yellow Transparent, which is so commonly cultivated as an early apple, is very seriously attacked by the disease. In the Eastern states, Missouri Pippin and York Imperial and to some extent the Ben Davis are badly attacked by blossom blight and twig blight at certain times and places. On the other hand, the Winesap group have proved quite resistant to the disease.

(2) The soil and the various other problems connected with its cultivation and fertilization exert an important influence. Fertilizers induce the trees to grow faster and therefore to blight more seriously. The use of artificial fertilizers and stable manure also stimulates the trees to blight. I believe commercial fertilizers are very seldom used in Colorado orchards. Stable manure is sometimes applied with the usual result of stimulating the trees into vigorous but tender growth, which is most susceptible to this malady. The soil moisture is an extremely important factor. Unless the soil is made too wet, soil moisture greatly favors the disease. Previous to investigating the blight in Colorado I was under the impression that, in handling orchards under irrigation, by withholding water or reducing it to the minimum amount necessary for sustaining the trees, blight could be practically checked. Apparently experience on the western slope especially goes against this theory. I have not, however, entirely given up the idea, and think still that in the irrigated orchards, where seriously attacked by blight or threatened by it, the amount of water should be reduced to the minimum

which is necessary for healthy but moderate growth and a fair crop of fruit. Cultivation not only makes plant food available, but causes the soils to retain their moisture in the lower layers, where the roots of the trees can obtain it. High culture, therefore, favors the blight. Unfortunately, we see that the conditions involving high-class cultivation and care of the orchard are those which favor the blight. It therefore may not be feasible to allow the blight question to govern us in the cultivation of the orchard. However, it should be seriously considered and moderation is advised.

(3) Humidity of the air exerts an important influence on the amount of moisture in the trees, and therefore upon the blight. When the trees are breathing off their moisture freely in a dry air and the condition is changed, the air becoming partly or wholly saturated, the tree immediately becomes gorged with water, and the conditions are favorable for the development of the disease. In the Eastern states atmospheric humidity and moist soil conditions commonly go together, for when it rains both the air and soil are moistened. However, in Colorado the two are distinct, the air being mostly dry, and the soil being under the control of the orchardist.

(4) The age of the tree is an important factor in the development of the disease. The worst period for blighting is the first five or six years after the trees come into blossoming and bearing. After that they become gradually less and less susceptible as they grow older. The growth of an old tree is slower and more conservative, and is not prolonged throughout the season as with a young one. Blight usually, therefore, only kills a few twigs or branches or blossom clusters. I have seen old trees, however, killed by pear blight, which runs into the base of the tree through watersprouts, and recently in Colorado I saw blight on apple trees ten inches or so in diameter, the disease having entered the base of the trees by some unknown method other than a watersprout.

(5) The style of pruning has an important influence on the amount of damage which the blight may do to the tree. It is also an important factor in determining the ease with which blight may be removed from the tree by cutting it out. The vase-form tree is undoubtedly the best as regards pear blight. This is true especially if the fruit spurs and fruiting branches are kept well out towards the end of the branches, and the branches themselves are

repeatedly forked. Pyramid trees may be rendered somewhat more resistant and convenient for handling if all the fruit spurs and small branches are pruned out of the center of the tree and kept at a distance from the main stem.

(6) The presence of other diseases in the Eastern states exerts an important influence on pear blight. The leaf blight and fungous diseases, as well as root-rot, crown-gall, and other troubles, weaken the tree, diminish its growth, or cause it to ripen up prematurely. These factors generally tend toward retarding the blight. Sometimes, however, diseased trees will blossom and catch the blight, while the healthy trees do not. I have seen trees which bloomed in the fall through the injurious effects of pear blight, the result being that the autumn bloom was attacked by pear blight, and, winter coming on shortly, the disease was able to hold over in such trees. The remarkable absence of leaf blights and other fungous diseases in Colorado are undoubtedly one of the factors which helps the pear blight to become more serious. The very health and vigor of these Colorado orchards render them susceptible to the disease.

(7) The state of nutrition of the tree also determines its relative immunity to blight. I have already indicated this in speaking of diseases, but it may be well to note that the conditions in Colorado are extremely favorable for assimilation. One of the principal foods of pear blight is the sugar as it exists the soft bast and tissues of the bark, twigs, etc. The high percentage of sunny days and intensity of sunshine in Colorado, the favorable temperature, along with the fertility of the soil, make conditions all favorable for assimilation of large amounts of elaborated sap. Pear trees and apple trees, therefore, in Colorado, are usually well fed or in a high state of nutrition. This is evidenced by the enormously large crops of fruit which they bear, and which usually are borne annually-a condition of affairs which is extremely rare in the humid regions of the eastern half of the United States. This high state of nutrition favors the blight.

SUMMARY OF TREATMENT

We may sum up the practical points in regard to treatment which may be gathered from the above discussion.

Use stable manure or other fertilizers in moderation or not at all.

In orchards threatened by destruction from pear blight, give the minimum amount of water necessary for the successful growth and cropping of the orchard.

Practice moderation in orchard cultivation.

Prune the trees in the vase form, if possible, or, at any rate, keep the fruit spurs, small branches and watersprouts away from the center of the tree.

Where it can be done without too much hardship, it is undoubtedly wise to remove bees during the blossoming time, so as to reduce the number of insect visitors to as small a number as possible.

Where practicable, plant varieties that are relatively immune to disease. As far as possible avoid planting non-resistant varieties of pears near apple orchards.

Above all, go over the orchards carefully in the fall, and re-inspect again in the spring, and cut out all cases of hold-over blight. I believe in cutting out pear blight in the summer time as well, but the main dependence should be placed upon the fall and winter cutting of the disease. While cutting out the blight, use a disinfectant made by dropping a tablet of corrosive sublimate into a pint of water. The tablet should be of sufficient strength to give a solution of one part corrosive sublimate to 1,000 parts water. Carry a small sponge or a wad of cloth and carefully wipe off the knives, saws or tools used in the cutting after the removal of a limb. As an extra precaution, it is also advisable to wipe off the freshly-cut surface with a sponge saturated with this solution.

Conclusion
Fire Blight in the 20th Century

The epochal discoveries of Burrill, Arthur, and Waite were just the beginning of a long line of research devoted to fire blight that continues unabated to this day. Those three pioneers established the causal agent of the disease and one of its most significant forms of dissemination, but left much to discover. Fire blight continued to be a frequently debilitating problem for orchardists throughout North America, and worse, as the 20th century progressed, the disease spread to at least 40 countries around the world where it was previously unknown. Researchers have made important discoveries concerning the nature of the pathogen and the disease, but fundamental mysteries still surround it. Given the fact that no "magic bullet" for disease control has ever been discovered, fire blight has spawned unique programs for its management in the field. Advances in molecular biology have opened new vistas on *E. amylovora*, but it remains to be seen when this will further the goal of controlling fire blight.

While Waite provided the evidence for the dissemination of *E. amylovora* by insects, research over the succeeding decades shed new light on the spread of this devastating disease. In 1917, Gossard and Walton demonstrated the importance of dissemination through water or mist. Ivanoff and Keitt showed in 1937 that wind, despite Waite's earlier assertion to the contrary, could play a greater role in pathogen spread than previously thought, particularly via the dissemination of bacterial strands.[1]

In terms of the geographic spread of fire blight, the 20th century was an active and sometimes nerve-wracking period. The disease moved inexorably around the globe, settling into hospitable niches from which it could not be removed. Although losses due to fire blight are often difficult to assess with accuracy, clearly pear and apple growers have suffered significant economic losses from the disease wherever it appeared.

The first epidemics of fire blight in California and the Pacific Northwest in the early 20th century were a specter of things to come. The disease soon moved from California up into Oregon and Washington. Then, in 1911, fire blight was reported in British Columbia in Canada.[2] The disease had been seen already in the eastern parts of Canada in the mid-19th century, much as it plagued the Hudson Valley and Midwest of the United States. It finally reached all parts of North America by 1921 when it was reported in the area around Mexico City.[3]

Outside of North America, a disease similar to fire blight was reported on apple in Japan in 1903.[4] Then, in 1919, the north island of New Zealand suffered an outbreak. Attempts to prevent the spread of the disease by quarantining both plant material and bees failed. Fire blight arrived on the south island by 1929.[5] New Zealand would soon respond by becoming a center for effective research on the disease.

Fire blight reached Europe later in the century. The United Kingdom experienced the first outbreaks in 1958 which damaged susceptible, locally popular pear varieties.[6] The disease took a few years to reach the Continent, but when it did, it spread rapidly, possibly due to migratory birds.[7] Fire blight was discovered in the Netherlands and Poland in 1966, two years later in Denmark, West Germany in 1971, and the next year in both Belgium and France.[8] Eradication attempts have been ineffective ultimately, but fortunately, the impact of fire blight tends to be limited in most years.[9] However, severe outbreaks do occur with relative frequency and they have important economic repercussions.

As fire blight spread across northern Europe, it was also advancing through the Mediterranean region. The disease appeared in Egypt in 1964, and by the late 1980s, nearly half of the pear trees in that country had been eradicated.[10] Fire blight was reported in Cyprus and Israel in the mid-1980s.[11] Then, it moved quickly into Turkey, Crete, Greece, and the Balkans.[12] It jumped to southern Italy, and by the mid-1990s, it had appeared in Spain.[13] There are few parts of the globe which are ecologically hospitable to the pathogen that have not become home to the disease.

It is not surprising that the geographical spread of fire blight throughout the 20th century fostered a great deal of research on the disease and the pathogen. Obviously, issues of disease control were extremely important and advances were made. Pruning, sanitation, and quarantine continued to be widely used. In addition, breeding for disease resistance was undertaken with great vigor. This program found some success at universities and experiment stations across the United States, and one large breeding program run out of the USDA research center in Beltsville, Maryland, can trace its roots back to Waite.[14] Later in the 20th century, some programs began to emphasize breeding for resistance in rootstocks. By the end of the century, scientists were using new molecular techniques to incorporate novel sources of resistance into varieties of apples and pears as well as rootstocks.[15]

In the early part of the 20th century, researchers searching for chemical help to control fire blight found limited success using copper sprays, particularly Bordeaux mixture. Waite had, in fact, provided some preliminary evidence of Bordeaux's effectiveness in his early work. However, copper's action as a preventative, combined with the frustratingly unpredictable nature of fire blight outbreaks, required growers to spray frequently in order to prevent infection, often resulting in russeting of fruit. This meant that copper sprays often were not practical for orchards producing fruit not intended for processing.

Then, in the 1950s, the antibiotic streptomycin came into practical use, replacing or sometimes complementing copper-based sprays. But in the 1970s, streptomycin-resistant strains of *E. amylovora* were found in California, Oregon, and Washington. Subsequent reports from Missouri and Michigan have raised concerns about the long term future of streptomycin as an effective control measure for fire blight and have fostered the search for alternatives. One alternative in the area of chemical control has been additional work late in the century involving chemicals, such as prohexadione calcium, which do not affect the bacteria directly but instead operates on the host to make it less susceptible in some fashion.[16]

The limits of chemical controls and the problems of resistance led to efforts to create a system that could predict the occurrence of the disease. A predictive system of this type would allow for a more focused and efficient use of chemicals. Therefore, the epidemiology of fire blight became a subject of intense focus in the 1970s and 1980s. One of the most fundamental contributions to this line of work was the research on the epiphytic nature of *E. amylovora* and the role of this surprisingly high population in the epidemiology of the disease.[17] It had long been understood that fire blight outbreaks depended on a seemingly endless parade of variables affecting different outbreaks, including the presence of cankers, changes in temperature, humidity, rain, as well as other weather factors such as damage from wind or frost or hail that might invite infection, the presence of insect vectors, and the presence of blight in neighboring orchards. Probably the best known and most widely distributed commercial assessment model was *Maryblyt* designed by Paul Steiner at the University of Maryland in 1990.[18]

Even as scientists were studying the role of the pathogen and disease in the environment, other researchers were digging deeper into the mysterious biology of the bacterium. *E. amylovora* had always appeared to be a monolithic species with an almost unbelievable level of homogeneity. While *E. amylovora* was easily separated from other bacteria, there did not seem to be any great differences within the species according to all the physiological and biochemical differentiation techniques available through the 1970s. That monolith began to break down in the 1980s with the appearance of new methods to pick apart strain differences including differential hosts, comparative phage studies, and genomics. This work on differentiation will have implications for disease control.[19]

The nature of the fire blight bacterium's pathogenicity obviously held special appeal to plant pathologists, and it turned out to be an important and frustrating area of research that has still not been settled to any certainty. Burrill made observations on the bacterium's degradation of starch in host cells that were clearly incorrect. Arthur searched for the bacterium's mechanism of damaging the host, a poison that he referred to as a "ptomaine," but without success. The bacterium's presence and its necrosis of host tissue had long been documented, but there were varying opinions as to exactly what the pathogen was doing to create the damage in the host tissue. Some researchers believed that the bacteria disrupted water flow by crowding the xylem vessels.[20] The bacterium's mode of attack on host cells was debated as well, with some researchers noting dissolved cell walls and others not. Mid-century applications of biochemistry to plant pathology primed researchers to search for toxins but with debatable results.

The "toxin" they found was part of the omnipresent ooze associated with fire blight infection. It became the subject of research using newly emerging genetic and molecular techniques in the 1970s. This research uncovered new information on the role of amylovoran, an exopolysaccharide necessary for pathogenicity, and levan, a homopolymer associated with virulence.[21] Still, this research did not settle issues of pathogenicity.

The search for pathogenicity moved into new realms with the discovery of *hrp* genes in *E. amylovora*.[22] These genes regulate the production of harpins, a protein associated with pathogenicity in susceptible hosts and the induction of a hypersensitive response in non-hosts. This has brought scientists even closer to an understanding of just how *E. amylovora* induces disease, and work on harpins and other chemicals has opened the possibility of using induced resistance for disease control.[23]

Aside from the molecular approach, other work on disease control has been fostered by recent research. Advances in the area of biological control have been made both in the laboratory and in the field. A biological control technique that has found some success with fire blight has been the use of antagonistic bacteria which compete with *E. amylovora* on the surface of flower parts and leaves and/or produce antibiotics. Interestingly enough, one of the antagonists being tested is *E. herbicola* (reclassified as *Pantoea agglomerans*), Waite's "false yellow germ."[24] However, as with many other diseases, practical success with biological control has been elusive.

All of the various control measures for fire blight are best used in combination to manage this difficult disease. Fire blight has become an excellent candidate for management programs that blend cultural controls, including resistant varieties, sanitation, and vegetative management, sprays of chemicals and biological control organisms, and disease forecasting. These integrated techniques have allowed growers not only to diminish the use of chemicals in their orchards, but bring hope for sustainable production of pears and apples even as the disease continues to spread around the world and encompass all regions that produce pome fruits.

Research continues on *E. amylovora* and fire blight and, no doubt, that research will continue to reveal fascinating aspects of this peculiar organism and make contributions to basic biology. This is much in the tradition of Burrill, Arthur, and Waite. These scientists made fundamental contributions to biology, but at the roots of their work was an interest in a plant disease and a mandate to ameliorate the devastating effects of that disease on important economic crops. Just as Burrill, Arthur, and Waite did in their day, the research that continues to this day on *E. amylovora* and fire blight will somehow find its way from the laboratory to the field and the best of it will find a useful role in agriculture.

LITERATURE CITED

1. Gossard and Walton 1917, 357-364; Ivanoff and Keitt 1937, 702-709.
2. Eastham 1935, 3229-3232.
3. Ramírez 1921, 662-663.
4. Uyeda 1903, 1-3.
5. Campbell 1920, 181-182.
6. Crosse, Bennett, and Garrett 1958, 1530; Lelliott 1959, 564-568.
7. Meijneke 1972, 128-136; van der Zwet 1994, 209-214.
8. Borecki, Basak, Zawadzka, and Millikan 1967, 3; Klarup 1969, 143-146; Meijneke 1972, 128-136; European and Mediterranean Plant Protection Organization (EPPO) 1972.
9. Klarup, 1969.
10. El-Helaly, Abo-El-Dahab, and El-Goorani 1964, 156-163.
11. Zutra and Shabi 1985, 151; Psallidas and Dimova 1986, 61-70.
12. European and Mediterranean Plant Protection Organization (EPPO) 1987; Psallidas 1990, 25-32.
13. Cariddi and Piglionica 1992, 265-266; de la Cruz Blanco 1996, 22-27.
14. van der Zwet and Keil 1979, 112-154; Lespinasse and Aldwinckle 2000, 253-273; Norelli and Aldwinckle 2000, 275-292.
15. Aldwinkle and van der Zwet 1979, 13-25; Aldwinkle, Norelli, Mills, and Brown 1996, 411; Aldwinkle, Ko, Norelli, Brown, and During 1998, 5, 3, 17.
16. Jones and Schnabel 2000, 235-251; Psallidas and Tsiantos 2000, 199-234.
17. Miller 1984, 63-68; Thomson 1986, 476-482; Thomson 1992, 1077.
18. Steiner 1990, 139-148; Lightner and Steiner 1993, 131-126.
19. Vanneste 1995, 21-41.
20. Sjulin and Beer 1977, 89-94.
21. Steinberger and Beer 1988, 135-144; Bernhard, Coplin, and Geider 1993, 158-168; Geier and Geider 1993, 387-404.
22. Bauer and Beer 1987, 169-170; Bauer and Beer 1991, 493-499; Steinberger and Beer 1988, 135-144.

23. Paulin 2000, 87-116.

24. Ishimaru, Klos, and Brubaker 1988, 746-750; Vanneste 1996, 67N-78N.

Bibliography

Aldwinkle, H. S. and T. van der Zwet. 1979. Recent progress in breeding for fire blight resistance in apples and pears in North America. *EPPO Bulletin* 9:13-25.

Aldwinkle, H. S., J. L. Norelli, J. Z. Mills, and S. K. Brown. 1996. Transformation of Gala apple with lytic protein genes. *Acta Horticulturae* 411:411.

Aldwinkle, H. S., K. Ko, J. L. Norelli, S. K. Brown, and K. During. 1998. Enhanced resistance to fire blight of apple lines transgenic for attacin E and T4 lysozyme genes. In *Seventh International Congress of Plant Pathology*. Vol. 3. Edinbugh: ICPP. 5, 3, 17.

Arthur, J. C. [1878] 1981. On the structure of the *Echinocystis lobata*. Reprint in *Iowa State Journal of Research* 55:219-234.

Arthur, J. C. 1885a. Diseases of the pear: Pear blight. In *Third Annual Report of the Board of Control of the New York Agricultural Experiment Station, for the Year 1884.* Albany: Weed, Parsons and Company, 357-367.

Arthur, J. C. 1885b. Pear blight and its cause. *American Naturalist* 19:1177-1185.

Arthur, J. C. 1885c. Proof that bacteria are the direct cause of the disease in trees known as pear blight. In *Proceedings of the American Association for the Advancement of Science, thirty-fourth meeting, held at Ann Arbor, Mich., August 1885.* Salem, MA: AAAS, 34:295.

Arthur, J. C. 1886. History and biology of fire blight. *Proceedings of the Philadelphia Academy of Natural Sciences* 38:332.

Arthur, J. C. 1887. Pear blight *Micrococcus amylovorus* (Bur.). In *Report of the Commissioner of Agriculture. 1886.* Washington, D.C.: GPO, 125-129.

Arthur, J. C. 1916. Cursory memorandum. *Personal History for Alumni Records, Iowa State College.* Arthur Herbarium. Department of Botany and Plant Pathology. Purdue University.

Baker, K. F. 1971. Fire blight of pome fruits: The genesis of the concept that bacteria can be pathogenic to plants. *Hilgardia* 40:603-633.

Barrett, J. T. 1918. Thomas J. Burrill (1839-1916). *Phytopathology* 8:1.

Bauer, D. W. and S. V. Beer. 1987. Cloning of *Erwinia amylovora* DNA involved in pathogenicity and induction of the hypersensitive reaction. *Acta Horticulturae* 217:169-170.

Bauer, D. W. and S. V. Beer. 1991. Further characterization of an *hrp* gene cluster of *Erwinia amylovora. Molecular Plant-Microbe Interactions* 4:493-499.

Beecher, H. W. 1844. The blight in the pear tree, its cause and a remedy for it. *The Magazine of Horticulture, Botany, and All Useful Discoveries and Improvements in Rural Affairs* 10:441-456.

Belden, H. 1837. Blight in pear trees. *Farmers' Register* 5:461.

Bernhard, F., D. L. Coplin, and K. Geider. 1993. A gene cluster for amylovoran synthesis in *Erwinia amylovora*: Characterization and relationship to *cps* genes in *Erwinia stewartii. Molecular General Genetics* 239:158-168.

Biographical Files, National Fungus Collection. U.S. Department of Agriculture Systematic Botany and Mycological Laboratory, Beltsville, MD.

"Blight in pear trees, etc." 1829. *New England Farmer* 8:57.

"The Blight in the pear tree." 1847.*The Horticulturist and Journal of Rural Art and Rural Taste* 1:59.

Bonn, W. G. and T. van der Zwet. 2000. Distribution and economic importance of fire blight. In *Fire Blight: The Disease and Its Causative Agent, Erwinia amylovora.* ed. by J. L. Vanneste. New York: CABI Publishing, 39.

Borecki, Z., W. Basak, B. Zawadzka, and D. Millikan. 1967. Fire blight in continental Europe. *Plant Disease Reporter* 51:3.

Brock, T. D. 1988. *Robert Koch: A Life in Medicine and Bacteriology.* Madison, WI: Science Tech Publishers.

Buel, J. 1827. The disease in pear and apple trees. *New England Farmer* 6:108.

Buel, J. 1832. Blight in pear trees. *Southern Agriculturist* 5:54.

Burrill, T. J. 1869. Agricultural botany. *Illinois Industrial University, minutes of the Board of Trustees, 2nd Report* 319-330.

Burrill, T. J. 1878. Report on botany and vegetable physiology: Pear blight. *Transactions of the Illinois State Horticultural Society for 1877* 11:114-116.

Burrill, T. J. 1879. Fire blight. *Transactions of the Illinois State Horticultural Society for 1878* 12:79-82.

Burrill, T. J. 1881a. Anthrax of fruit trees; or the so-called fire blight of pear, and twig blight of apple trees. In *Proceedings of the American Association for the Advancement of Science, twenty-ninth meeting held at Boston, Mass. August, 1880.* Salem, MA: AAAS, 29:583-597.

Burrill, T. J. 1881b. Bacteria as a cause of disease in plants. *American Naturalist* 15:527-531.

Burrill, T. J. 1881c. Blight. *Botanical Gazette* 6:271-273.

Burrill, T. J. 1881d. Blight of pear and apple trees. In *Illinois Industrial University Board of Trustees Tenth Report for Two Years, ending August, 1880.* Springfield, IL: H.W. Rokker, 62-84.

Burrill, T. J. 1881e. Report on botany and vegetable physiology: pear and apple-tree blight. *Transactions of the Illinois State Horticultural Society for 1880* 14:157-167.

Burrill, T. J. 1882. The bacteria: An account of their nature and effects, together with a systematic description of the species. In *Illinois Industrial University Board of Trustees Eleventh Report.* Urbana: Illinois Industrial University, 93-157.

Burrill, T. J. 1883a. New species of *Micrococcus* (Bacteria). *American Naturalist* 17:319-320.

Burrill, T. J. 1883b. Pear blight and peach yellows. *Transactions of the Illinois State Horticultural Society* 17:46-50.

Burrill, T. J. 1887. A disease of broom-corn and sorghum. In *Proceedings of the Eighth Annual Meeting of the Society for the Promotion of Agricultural Science*, ed. by W. R. Lazenby, Columbus, OH: Gazette Printing House, 30-36.

Burrill, T. J. 1889. A bacterial disease of Indian corn. In *Proceedings of the Tenth Annual Meeting of the Society for the Promotion of Agricultural Science*, ed. by W.R. Lazenby, Columbus, OH: Gazette Printing House, 19-25.

C. D. 1844. On the blight of pear trees. *Western Farmer* 4:275.

Campbell, C. L., P. D. Peterson, and C. S. Griffith. 1999. *The Formative Years of Plant Pathology in the United States*. St. Paul, MN: American Phytopathological Society.

Campbell, J. A. 1920. The orchard: The outbreak of fire blight. *New Zealand Journal of Agriculture* 20:181-182.

Campbell, R. N. 1979. Fire blight. *Natural History* 88:62-69.

Cariddi, C. and V. Piglionica. 1992. First record of fire blight in Italy. *Phytoparasitica* 19(31):265-266.

Castor, J. Y. 1839. Blast in pear trees. *Western Farmer* 1:8.

Coleman, W. R. 1977. *Biology in the Nineteenth Century: Problems of Form, Function, and Transformation*. Cambridge, MA: Cambridge University Press.

Coxe, W. 1817. *A View of the Cultivation of Fruit Trees, and the Management of Orchards and Cider; with Accurate Descriptions of the Most Estimable Varieties of Native and Foreign Apples, Pears, Peaches, Plums, and Cherries, Cultivated in the Middle States of America*. Philadelphia: M. Carey and Son.

Crosse, J. E., M. Bennett, and C. M. E. Garrett. 1958. Fire blight of pear in England. *Nature* 182:1530.

Cummins, G. B. 1943. Joseph Charles Arthur (1850-1942). *The American Philosophical Society Year Book 1942*. Philadelphia: American Philosophical Society, 328-331.

Cummins, G. B. 1978. J. C. Arthur: The man and his work. *Annual Review of Phytopathology* 16:20-21.

"Cure for blight of pear trees." 1836-1837. *Farmers' Register* 4:395.

de la Cruz Blanco, J. 1996. Lucidencias climaticas y fytosanitarias en los cultivos espanoles durante 1995. *Phyto-España* 78:22-27.

Denning, W. 1794. On the decay of apple trees. *Transactions of the New York Society for the Promotion of Agriculture, Arts and Manufacturers* 2:219-222.

Downing, A. J. 1872. *The Fruits and Fruit-trees of America; or, the Culture, Propagation, and Management, in the Garden and Orchard, of Fruit-trees Generally; with Descriptions of all the Finest Varieties of Fruit, Native and Foreign, Cultivated in this Country*. New York: John Wiley and Son.

Eastham, J. W. 1935. The relation between climate and the incidence of some orchard diseases in British Columbia. *Pacific Science Congress Proceedings* 5:3229-3232.

El-Helaly, A. F., M. K. Abo-El-Dahab, and M. A. El-Goorani. 1964. The occurrence of the fire blight disease of pear in Egypt. *Phytopathologia Mediterranea* 3:156-163.

Elliot, C. W. 1844. The blight. *Western Farmer* 4:255.

European and Mediterranean Plant Protection Organization (EPPO). 1972. The Discovery of Fire Blight in Northern France. *European and Mediterranean Plant Protection Organization. Report 366.* Paris.

European and Mediterranean Plant Protection Organization (EPPO). 1987. Fire Blight Outbreaks in Certain Greek Islands. *European and Mediterranean Plant Protection Organization. Report 476.* Paris.

Fisher, D. V., and W. H. Upshall, eds. 1976. *History of Fruit Growing and Handling in United States of America and Canada, 1860-1972.* Kelowna, British Columbia: Regatta City Press.

Forsyth, W. 1802. *A Treatise on the Culture and Management of Fruit Trees; in Which a New Method of Pruning and Training is Fully Described, Together with Observations on the Diseases, Defects, and Injuries, in all Kinds of Fruit and Forest Trees; as also, an Account of a Particular Method of Cure, Made Public by Order of the British Government, to Which are Added, an Introduction and Notes, Adapting the Rules of the Treatise to the Climates and Seasons of the United States of America by William Cobbett.* Philadelphia: J. Morgan.

G. S. B. [Gookins, S. B.] 1846. Remarks on pear blight of the west. *Horticulturist* 1:254-255.

Galloway, B. T. 1892. Report of the chief of the Division of Vegetable Pathology. In *Report of the Secretary of Agriculture, 1891.* Washington, D.C.: GPO.

Galloway, B. T. 1900. Progress in the treatment of plant diseases in the United States. In *U.S. Department of Agriculture Yearbook 1899.* Washington, D.C.: GPO.

Geier, G. and K. Geider. 1993. Characterization and influence on virulence of the levansucrase gene from the fire blight pathogen *Erwinia amylovora. Physiological Molecular Plant Pathology* 42:387-404.

Glawe, D. A. 1992. Thomas J. Burrill, pioneer in plant pathology. *Annual Review of Phytopathology* 30:17-18.

Gossard, H. A. and R. C. Walton. 1917. Fire-Blight Infection: Rain proves an important carrier of disease in open orchards. *Ohio Agricultural Experiment Station Monthly Bulletin 2*. Wooster, OH: Ohio Agricultural Experiment Station, 357-364.

Hedrick, U. P. 1950. *A History of Horticulture in America to 1860*. New York: Oxford University Press.

Hottes, C. F. 1940. Personal recollections of Thomas J. Burrill and his work. *Illinois Alumni News* 6-7.

Hull, E. S. 1869. Cryptogamous diseases. *Transactions of the Illinois State Horticultural Society* 2:35.

Ishimaru, C. A., E. J. Klos, and R. R. Brubaker. 1988. Multiple antibiotic production by *Erwinia herbicola*. *Phytopathology* 78:746-750.

Ivanoff, S. S. and G. W. Keitt. 1937. The occurrence of aerial bacterial strands on blossoms, fruits, and shoots blighted by *Erwinia amylovora*. *Phytopathology* 27:702-709.

Jones, A. L. and E. L. Schnabel. 2000. The development of Streptomycin-resistant strains of *Erwinia amylovora*. In *Fire Blight: The Disease and its Causative Agent, Erwinia amylovora*. ed. by J. L. Vanneste. New York: CABI Publishing, 235-251.

Kennedy, B. F., K. D. Widkin, and K. F. Baker. 1979. Bacteria as the cause of disease in plants: A historical perspective. *ASM News* 45:2.

Kenrick, W. 1841. *The New America Orchardist: or an Account of the Most Valuable Varieties of Culture, Management, Uses, etc. with an Appendix on Vegetables, Ornamental Trees, Shrubs, and Flowers, the Agricultural Resources of America, and on Silk, etc.* 3rd ed. Boston: Otis, Broaders, and Co.

Kern, F. D. 1942. Obituary: Joseph Charles Arthur. *Science* 95:617.

Klarup, J. 1969. Fire blight'en pa Nordfalster. *Erhvervsfrugtavleren* 36:143 146.

Lazell, J. A. 1840. Observations upon the blight in fruit trees, with an account of a method of preventing the disease. *The Magazine of Horticulture* 6:415.

Lelliott, R. A. 1959. Fire blight of pears in England. *Agriculture* 65:564-568.

Lespinasse, Y. and H. S. Aldwinckle. 2000. Breeding for resistance to fire blight. In *Fire Blight: The Disease and its Causative Agent, Erwinia amylovora*. ed. by J. L. Vanneste. New York: CABI Publishing, 253-273.

Lightner, G. W. and P. W. Steiner. 1993. An update on version 4.1 of the Maryblyt computer program for predicting fire blight. *Acta Horticulturae* 338:131-126.

Logue, J. N. 1995. *Beyond the Germ Theory: The Story of Dr. Cooper Curtice.* College Station, TX: Texas A&M University Press.

Lowell, J. 1826. Insect in pear trees. *New England Farmer* 5:17-18.

Magner, L. N. 1994. *A History of the Life Sciences.* New York: Marcel Dekker, Inc.

Mains, E. B. 1942. Joseph Charles Arthur (1850-1942). *Mycologia* 34:601 605.

McGregor, S. E. 1976. *Insect Pollination of Cultivated Crop Plants.* USDA Agricultural Handbook 496. Washington, D.C.: GPO.

McMahon, B. 1806. *The American Gardener's Calendar; Adapted to the Climates and Seasons of the United States.* Philadelphia: B. Graves.

Meijneke, C. A. R. 1972. Perevuur en zijn verspreiding. *Gewasbescherming* 3:128-136.

Miller, H. J. 1984. *Erwinia amylovora* detection and its significance in survival studies. *Acta Horticulturae* 151:63-68.

Moore, V. A. 1894. An inquiry into the alleged relation existing between the Burrill disease of corn and the so-called corn-stalk disease of cattle. *Agricultural Science* 8:368-369.

Norelli, J. L. and H. S. Aldwinckle. 2000. Transgenic varieties and rootstocks resistant to fire blight. In *Fire Blight: The Disease and Its Causative Agent, Erwinia amylovora.* ed. by J. L. Vanneste. New York: CABI Publishing, 275-292.

Overman, C. R. 1855. Remarks on the pear and its culture. *Transactions of the Illinois State Agricultural Society: With the Proceedings of the County Societies, and Kindred Associations* 1:447-451.

Paulin, J.-P. 2000. *Erwinia amylovora*: General characteristics, biochemistry and serology. In *Fire Blight: The Disease and Its Causative Agent, Erwinia amylovora.* ed. by J. L. Vanneste. New York: CABI Publishing, 87-116.

Peffer, G. 1882. Apple-tree blossoms-blight. *Transactions of the Wisconsin State Horticultural Society* 12:191-202.

Peterson, P. D. and C. L. Campbell. 1997. Beverly T. Galloway: visionary administrator. *Annual Review of Phytopathology* 35:29-43.

Psallidas, P. G. 1990. Fire blight of pomaceous trees in Greece – evolution of the disease and characteristics of the pathogen *Erwinia amylovora. Acta Horticulturae* 273:25-32.

Psallidas, P. G. and M. Dimova. 1986. Occurrence of the disease fire blight of pomaceous trees in Cyprus – characteristics of the pathogen *Erwinia amylovora. Annals of the Institute of Phytopathology* 15:61-70.

Psallidas, P. G. and J. Tsiantos. 2000. Chemical control of fire blight. In *Fire Blight: The Diseases and Its Causative Agent, Erwinia amylovora.* ed. by J. L. Vanneste. New York: CABI Publishing, 199-234.

Ramírez, R. 1921. Plagas de la agricultura en el distrito federal. *La Revista Agrícola* 9:662-663.

Reed, G. 1943. Phytopathology 1867-1942. *Torreya* 43:162-163.

Roberts, J. W. 1946. Merton Benway Waite. *Phytopathology* 36:175.

Rodgers, A. D. 1952. *Erwin Frink Smith: A Story of North American Plant Pathology.* Memoirs of the American Philosophical Society held at Philadelphia for promoting useful knowledge 31. Philadelphia: American Philosophical Society.

Rosenberg, C. E. 1997. No Other Gods: On Science and American Social Thought. Baltimore: Johns Hopkins University Press.

Salisbury, J. H. and C. B. Salisbury. 1864. Microscopic researches: Resulting in the discovery of what appears to be the cause of the so-called blight in the apple, pear and quince trees; and the decay in their fruit, &c. In *Eighteenth Annual Report of the Ohio State Board of Agriculture.* Columbus, OH: Richard Nevins, State Printer.

Sijulin, T. M. and S. V. Beer. 1978. Mechanism of wilt induction by amylovorin in cotoneaster shoots and its relation to wilting of shoots infected by *Erwinia amylovora. Phytopathology* 68:89-94.

Smith, E. F. 1911. *Bacteria in Relation to Plant Diseases.* Vol. 2. Washington, D. C.: Carnegie Institution.

Smith, E. F. 1920. *An Introduction to Bacterial Diseases of Plants.* Philadelphia: W.B. Saunders Company.

Smith, E. F. Papers. 1865–1940. American Philosophical Society. Philadelphia.

Solberg, W. U. 1968. *The University of Illinois 1867-1894: An Intellectual and Cultural History.* Urbana: University of Illinois Press.

Steinberger, E. M. and S. V. Beer. 1988. Creation and complementation of pathogenicity of mutants of *Erwinia amylovora. Molecular Plant-Microbe Interactions* 1:135-144.

Steiner, P. W. 1990. Predicting apple blossom infections by *Erwinia amylovora* using the Maryblyt model. *Acta Horticulturae* 273:139-148.

Thacher, J. 1822. *The American Orchardists; or a Practical Treatise on the Culture and Management of Apple and Other Fruit Trees, with Observations on the Diseases to Which They are Liable, and Their Remedies. To Which is Added the Most Approved Method of Manufacturing and Preserving Cider.* Boston: Joseph W. Ingraham.

Thomson, S. V. 1986. The role of the stigma in fire blight infections. *Phytopathology* 76:476-482.

Thomson, S. V. 1992. Monitoring flowers for presence of epiphytic *Erwinia amylovora* by stigma streaking. *Phytopathology* 82:1077.

Thornberry, H. H. Papers. 1929–1994. University of Illinois Archives. Urbana.

Thwaites, R. G., ed. 1900. *The University of Wisconsin. Its History and Its Alumni With Historical and Descriptive Sketches of Madison.* Madison, WI: J. N. Purcell.

Trelease, W. 1925. The evolution of a botanist and of a department of botany. *Proceedings of the Indiana Academy of Science* 34:59-62.

True, A. C. 1937. *A History of Agricultural Experimentation and Research in the United States, 1607-1925.* USDA Misc. Pub 251. Washington, D.C.: GPO.

USDA Letters Received (General), 1891-1900; 1901-1902; 1903-1908. Records of the Division of Vegetable Pathological and Physiological Investigations. Records of the Bureau of Plant Industry, Soils, and Agricultural Engineering. Record Group 54. National Archives, Washington, D.C.

USDA Letters Sent, 1885-1888. Records of the Division of Vegetable Pathological and Physiological Investigations. Records of the Bureau of Plant Industry, Soils, and Agricultural Engineering. Record Group 54. National Archives, Washington, D.C.

Uyeda, E. 1903. The causal bacterium of apple blight. *Dai-Nippon No-Kaiho* 260:1-3.

van der Zwet, T. 1994. The various means of dissemination of the fire blight bacterium *E. amylovora. European and Mediterranean Plant Protection Organization Bulletin* 24:209-214.

van der Zwet, T. and H. L. Keil 1979. *Fire Blight: A Bacterial Disease of Rosaceous Plants.* USDA Agricultural Handbook 510. Washington, D.C.: GPO.

Vanneste, J. L. 1995. *Erwinia amylovora.* In *Pathogenesis and Host Specificity in Plant Diseases: Histopathological, Biochemical, Genetic and Molecular Bases.* Vol. *Prokaryotes.* eds. Singh, U. S., R. P. Singh, and K. Kohmoto. Oxford: Pergamon Press. 21-41.

Vanneste, J. L. 1996. Honey bees and epiphytic bacteria to control fire blight, a bacterial disease of apple and pears. *Biocontrol News and Information.* 17:67N-78N.

Vanneste, J. L. 2000. What is fire blight? Who is *Erwinia amylovora*? How to control it. In *Fire Blight: The Disease and Its Causative Agent, Erwinia amylovora.* ed. by J. L. Vanneste. New York: CABI Publishing, 1-6.

Waite, M. B. 1891. Results from recent investigations in pear blight. *Botanical Gazette* 16:259.

Waite, M. B. 1892a. Results from recent investigations in pear blight. In *Proceedings of the American Association for the Advancement of Science. For the fortieth meeting, held at Washington, D.C. August 1891.* Salem, MA: AAAS, 315.

Waite, M. B. 1892b. The pear blight microbe. In *Transactions of the Peninsula Horticultural Society, fifth annual session held at Dover, Delaware. January 19th , 20th, and 21st, 1892.* Wilmington: A. P. Whitaker, Printer, 33-34.

Waite, M. B. 1894. *The Pollination of Pear Flowers.* USDA Division of Vegetable Pathology Bulletin No. 5. Washington, D.C.: GPO.

Waite, M. B. 1895. Pear blight. *Proceedings of the annual meeting of the Western New York Horticultural Society* 40:96-99.

Waite, M. B. 1896. The cause and prevention of pear blight. In *Yearbook of the United States Department of Agriculture. 1895.* Washington, D.C.: GPO, 295-300.

Waite, M. B. 1898a. The life-history and characteristics of the pear-blight germ. In *Proceedings of the American Association for the Advancement of Science, Forty-seventh meeting and fiftieth anniversary. Held at Boston, Mass. August, 1898.* Salem, MA: AAAS, 427-428.

Waite, M. B. 1898b. Pear blight and its treatment. *Annual Report of the New York State Agricultural Society* 57:785.

Waite, M. B. 1902. Relation of bees to the orchard. In *Proceedings of the Twenty-seventh session of the American Pomological Society held in Buffalo, NY, Sept. 12-13, 1901.* Lansing: American Pomological Society, 66-70.

Waters, J. 1839. Blight in fruit trees. *Western Farmer* 1:37.

Wendell, H. 1850. Report of Dr. Herman Wendell, of Albany: Pears. In *Report of the Commissioner of Patents for the year 1849. Part 2. Agriculture.* Washington, D.C.: GPO, 446-448

Woods, A. F. 1945. Merton B. Waite. *Science* 101:632

Zutra, D. and E. Shabi. 1985. First evidence of fire blight in Israel. *Phytoparasitica* 13:151.